Learning Science

LEARNING
Science

The Value of Crafting Engagement in Science Environments

Barbara Schneider, Joseph Krajcik,
Jari Lavonen, and Katariina Salmela-Aro

With a Foreword by Margaret J. Geller

Yale

UNIVERSITY PRESS

New Haven and London

Yale University Press books may be purchased in quantity for educational, business, or promotional use. For information, please e-mail sales.press@yale.edu (U.S. office) or sales@yaleup.co.uk (U.K. office).

Set in Scala Sans type by Integrated Publishing Solutions.
Printed in the United States of America.

Library of Congress Control Number: 2019941028
ISBN 978-0-300-22738-3 (hardcover : alk. paper)

A catalogue record for this book is available from the British Library.

This paper meets the requirements of ANSI/NISO Z39.48-1992 (Permanence of Paper).

10 9 8 7 6 5 4 3 2 1

Contents

Foreword

Look up at the dark night sky and wonder: How do stars shine? How big is the universe? So many people ask these questions but lack the basic tools to explore what we know and how we find answers to these profound inquiries about the universe.

Learning Science is a roadmap for encouraging wonder and discovery in the high-school classroom. It describes how two countries with very different cultures, Finland and the United States, have joined together in a common goal: the humanization of high-school science instruction.

Project-based learning is the fundamental concept behind the roadmap. Like any path-breaking scientific investigation, project-based learning units begin with an important question. Then, in a departure from the standard lecture format, they build inclusive discussion and hands-on discovery sessions. The interpersonal involvement heightens both student and faculty interest. *Learning Science* highlights the relative strengths and weaknesses of the American and Finnish education systems. Among other differences, the Finnish respect for teachers stands in stark contrast with the low esteem granted to U.S. teachers, despite their commitment to their profession.

Project-based learning offers a richer, more creative classroom environment for teachers and students. It seems possible that widespread use of these techniques could spur a return to the kind of professionalism and respect so important for success in any educational setting.

The United States offers Finland some insights into the richness and challenges of a diverse classroom. Many U.S. teachers are skilled in welcoming students from a wide variety of life experiences. This challenge is relatively new to Finland, which has experienced a recent surge in immigration.

Over my lifetime, there have been many new ideas about revising science curricula and encouraging students to enjoy these subjects and even to pursue careers in them. Most of these endeavors have been well-meaning failures, in part for lack of effective ways of evaluating their impact. *Learning Science* describes the remarkable cell phone surveys of both students and teachers that provide nearly real-time evaluation of the learning experience, including the all-important emotional responses on both sides.

Learning Science is a call to join an important experiment where we all wonder together. It encourages us all to vocalize those nagging questions. It inspires us to share the questions we ask every day with our children and with our friends and colleagues. It motivates us to keep asking until we find routes to the answers. In a world where numeracy and the understanding of basic scientific concepts are increasingly important, the project-based learning approach offers a richer way to solve these problems as a worldwide community of curious people who wonder.

Margaret J. Geller
Harvard-Smithsonian Center for Astrophysics
Cambridge, Massachusetts

Preface

Why do so many students find high-school science classes uninteresting? Why do they become easily disengaged when learning about the periodic table or Newton's second law? Declining interest and engagement in science is a worldwide issue, even in countries like Finland where students do well on international assessments. Student disenchantment with science is a real concern; it does not bode well for the future preparation and development of a scientific workforce or our global society's willingness to grapple with the menacing problems of environmental change, new human diseases, and the boundaries of artificial intelligence and robotics. Why is science, especially in high school, viewed by many as a boring, "have to take" subject? Is there something about the way that science is taught that contributes to the malaise that young people feel when thinking about their science classes? Young people love *Star Wars* and a multitude of other science-related fantasies, but in science class they spend their time daydreaming about anything but the very subject matter that might help make those fantasies a reality.

The scientific and policy community have recognized this problem and called for major reforms in science learning and instruction across the U.S. education system. Two influential documents, the *Framework for K–12 Science Education* and *Next Generation Science Standards* (NGSS), describe these reforms and outline a new vision of scientific proficiency that encourages students to make sense of phenomena and find solutions to problems using three-dimensional learning— which includes making use of disciplinary core ideas, scientific and engineering practices, and crosscutting concepts. The goal is for students to develop a deep understanding of scientific ideas and scientific practices, and to step away from rote memorization of science facts. This call for science reform has been embraced globally, including in Finland, where although students had high achievement in science, their engagement and interest in science was among the lowest of many industrialized countries. To address this problem, Finland has adopted Core Curriculum guidelines that are similar in focus and content to those in the United States.

Although the NGSS and the Finnish Core Curriculum guidelines identified what students should know and be able to do, they did not prescribe a specific curriculum or instructional ideas—the assumption was that curricula would need to be developed to enable students to think and use the practices of scientists and engineers (that is, to explain phenomena and design solutions). By not proposing specific curricular content, the NGSS opened the door for educators to reexamine, critique, and generate science experiences that would both engage students and meet the new standards. One instructional method that aligns well with the NGSS and new Finnish standards—and addresses the challenges facing both countries—is project-based learning (PBL), an approach that encourages students to use their imaginations and "figure out" phenomena by solving problems using scientific ideas and practices.

In both the United States and Finland, PBL is oriented toward teaching students the foundations of scientific research. Though the studies use the same designs, instruments, and analytic plans in both countries, certain ideas and their interpretations are slightly different—in part because of culture and language. The Finnish model of PBL emphasizes students' active role in constructing cognitive artifacts, like scientific relationships and models; group discussions and synthesis of ideas; engagement in scientific practices similar to those of professional scientists; development of understandings through collaborating or sharing, using, and debating ideas with others; and the use of cognitive tools, such as graphs, which help students to see patterns in data. Both countries underscore the importance of promoting engagement with real scientific practices and giving all students (both females and males) a more realistic idea of what career scientists do.

In 2015, the National Science Foundation and the Academy of Finland funded an international study to increase student engagement and achievement in high-school physics and chemistry. This study, entitled Crafting Engagement in Science Environments (CESE), brought together an unusual collaborative team of psychologists, sociologists, learning scientists, science educators, and teachers at Michigan State University and the University of Helsinki to conduct an empirical study to test the effect of PBL on how, and how well, students learn science, and what their academic, social, and emotional experience is like in the classroom. The U.S. team completed a field test of three physics and three chemistry science units in Michigan high schools (complete with unit assessments and instructional materials). Participating teachers received ongoing professional and technical support for implementing and refining the intervention. In addition, a dozen Finnish secondary-school teachers, who have also participated in U.S. professional learning activities, modified and used their own PBL materials and tested their impact. Through these efforts and work in several other countries, we are developing an international professional community focused on teaching and learning science.

To measure student engagement and social and emotional states, our project uses the experience sampling method (ESM). Students are signaled several times during the day and prompted to answer on their smartphones a short survey about where they are, what they are doing, and who they are with, as well as a number of questions related to their feelings at the moment when signaled. Science teachers involved in the project answer their own ESM surveys at the same time as their students. The participating teachers and students also complete surveys about their attitudes, background, academic well-being, and career goals. Classroom observations, video data, and student artifacts are also collected to measure fidelity of implementation, while pre- and post-tests measure student science learning.

To test the effectiveness of PBL, we have used a single-case design that requires the repeated application of the intervention between specified times. The data are collected from the students during business-as-usual instructional periods and then during the PBL intervention units. This pattern is replicated several weeks later for the same amount of time in both conditions. While single-case designs are appropriate for this stage of our study, in the next phase of our work we will be scaling up, conducting a cluster-randomized trial in multiple U.S. states and Finland that will reach thousands of students.

With the data we have collected thus far and the work we have published and presented at scientific conferences, we can state confidently that our results consistently show PBL's positive effects on students' social, emotional, and cognitive learning. Further, not only are the students experiencing positive effects from PBL, but their teachers also report (and we have been able to document) that PBL has transformed their practice. Some might say this is a result of selection bias: that teachers who are inclined to be innovative are drawn to this type of study. This could certainly be the case. Even so, the process of transformation that these teachers are experiencing— and the positive effects it is having on their students—are promising findings that should be shared as we move toward making major reforms in science education.

A special focus of this project is to bring high-quality science instruction to schools serving predominately low-income and minority students. Consequently, future evaluation efforts will consider both the average effect of the intervention and its differential effect on student subgroups. We have worked with more than sixty teachers in twenty-four schools and served 1,700 students in the United States and Finland, over a third of whom are low-income and/or minority. The project has generated several major papers and presentations, and the PBL units are currently being formatted for distribution by the Organisation for Economic Co-operation and Development (OECD) as part of its new initiative to develop teaching models that enhance creativity and critical thinking. Curriculum materials will also be available in the future at Creative Commons Open Source. All of our data are being archived at the Inter-university Consortium for Political and Social Research (ICPSR) and will be available for further research and replication. Data from this project will be available at doi.org/10.3886/E100380V1.

Another important component of our work is establishing a professional learning community among the teachers who participate in our project. Building this community begins with professional learning sessions in which teachers work with science educators to gain a better understanding of the NGSS and PBL—and continues as teachers share ideas and experiences and receive feedback in regularly held videoconferences. Several of these virtual sessions have included our Finnish team, so that participating U.S. teachers have had the opportunity to work with and learn from science teachers and university professors in Finland. This venue allows participating teachers to broaden their professional network and gain valuable insight into how instruction and curricular reforms are shaped by context.

Each year, we assemble a delegation of teacher interns, teachers, and teacher educators to travel to Finland as part of an international exchange. During these exchanges, participants learn about the Finnish education system, visit Finnish schools, and interact with teacher educators. The delegation for these exchanges consists of teachers in our project, undergraduate pre-service teachers, and science administrators. Over the past three years we have been fortunate to have the following individuals participate in these exchanges: Cameron Cochran, James Emmerling, Sandra Erwin, Morgan Gilliam, Kimberly Herd, Claire Jackowski, Jonathan Kremer, Lindsey Montayne, Megan O'Donovan, Will Paddock, Donna Pohl, Brandon Rubin, Kathryn Schwartz, and Lindsay Young. Their observations have sharpened our thinking about what we are learning about the Finnish education system, especially regarding teacher training and what Finland is learning about the U.S. system. A similar exchange has also occurred involving Finnish teachers, graduate students, and postdoctoral fellows who have visited the United States and observed in PBL classrooms. Finnish visitors over the past three years have included Johanna Jauhiainen, Timo Kärkkäinen, Tiia Karpin, Hilkka Koljonen-Toppila, Eija Kujansuu, Taina Makkonen, Julia Moeller, Ari Myllyviita, Annina Rostila, Krista Sormunen, Pauliina Toivonen, Hannes Vieth, and Panu Viitanen.

This is a huge study spanning two countries and there are multiple individuals to whom we owe special acknowledgments and thank yous. To begin with are the researchers who joined the project at the start and helped with the initial conceptual framework and design. They have moved on with their careers, but we have not forgotten their much-appreciated contributions. We thank, in Finland, Jukka Marjanen, Finnish Education Evaluation Centre (FINEEC); Julia Moeller, assistant professor, University of Leipzig; and Jaana Viljaranta, assistant professor, University of Eastern Finland. And in the United States we acknowledge Michael Broda, assistant professor, Virginia Commonwealth University; Justin Bruner, data analyst, Michigan State University; Jason Burns, research associate, Michigan State University; and Justina Spicer, education research analyst and consultant.

In developing our curricular approach, we relied on the expertise of several people. Certainly among the most important are the teachers who acted as lead developers and helped with the delivery of professional learning, online training, and support. Their assistance has been invaluable. We thank, in the United States, physics teachers Steve Barry, Cameron Cochran, John Plough, and Brandon Rubin, and chemistry teachers Sandra Erwin, Lindsey Montayne, and Will Paddock. In Finland, we thank Johanna Jauhiainen, Tea Kantola, Timo Kärkkäinen, Tiia Karpin, Katrin Kirm, Hilkka Koljonen-Toppila, Eija Kujansuu, Sanna Lehtamo, Taina Makkonen, Ari Myllyviita, Mikko Rahikka, Annina Rostila, Antero Saarnio, Krista Sormunen, Pauliina Toivonen, Simo Tolvanen, Virpi Vatanen, Hannes Vieth, Panu Viitanen, and Miikka de Vocht. (Finnish curriculum development tends to be a more widely shared activity than in the United States.) Additionally, to ensure that the PBL curriculum and instruction exemplified three-dimensional learning, we received critiques and suggestions from Melanie Cooper, Lappan-Phillips chair of science education and professor of chemistry, Michigan State University; David Fortus, associate professor, Department of Science Teaching, Weizmann Institute of Science, in Israel; Knut Neumann, professor, Leibniz Institute for Science and Mathematics Education at the University of Kiel (IPN), in Germany; and Ryan Stowe, postdoctoral researcher, Michigan State University. Their expertise and suggested improvements helped us keep our work true to the principles of the NGSS—we owe you all a special thank you. Additionally, this work has been occurring at the same time that PBL units are being designed and implemented in another study at Michigan State University on "Multiple Literacies in Project-Based Learning," funded by the George Lucas Educational Foundation. Many of that project's investigators, graduate students, postdoctoral fellows, and consultants conferred and offered advice on our work. We thank them for their contributions.

A project that delivers instruction and simultaneously collects data on it requires the skills of many people. We owe a special thank you to the undergraduates who, throughout the first three years of this project, worked with us to program smartphones and to distribute and retrieve the phones, and the materials needed for the units, at the field sites. Thanks to Garrett Amstutz, Jacob Herwaldt, Peter Hulett, Parker LaVanway, Elizabeth Paulson, Matthew Pottebaum, Hannah Weatherford, and Jacob Webb for fearlessly and tirelessly driving the wintry roads of Michigan. And a thank you to our graduate students, Megan O'Donovan and Lindsay Young, and our postdoctoral fellow I-Chien Chen, who have helped to develop the data files, code books, technical reports, and data analyses.

A special thanks to our advisory team, which has helped in planning for our larger study and given us thoughtful comments on our current data collection plans and analyses. Finnish and international participants include Risto Hotulainen, associate professor, University of Helsinki; Sari Lindblom, professor and vice rector, University of Helsinki; Kirsti Lonka, professor, University of Helsinki; Kirsi Tirri, professor, University of Helsinki; Risto Vilkko, docent, University of Helsinki, and program manager, Academy of Finland; and Stephan Vincent-Lancrin, deputy head of division and senior analyst at the OECD. And in the United States: Nhu Do, principal, Washtenaw International High School; Larry V. Hedges, board of trustees and professor of statistics, Northwestern University; and James Pellegrino, distinguished professor of Liberal Arts and Sciences, University of Illinois at Chicago. Another adviser to our work, whose open-source software design allowed us to collect high-quality and secure data, is Robert Evans, senior software engineer at Google. Thank you, Bob, for making much of our work on social and emotional learning possible.

Taking a walk through the universe with Dr. Margaret Geller, an astrophysicist at the Harvard-Smithsonian Center for Astrophysics, was a life-changing experience. Geller was remarkably willing to read our book—by researchers in science education, sociology, and social psychology whom she did not know—and write the foreword. We thank her deeply and hope to share additional explorations of our universe with her, and many more students and teachers, as we create new units in astrophysics.

Then there are the many individuals in the John A. Hannah and CREATE for STEM offices that helped with logistics, budgets, and university regulations. Special recognition goes to Juliana Brownrigg, Sue Carpenter, Michelle Chester, Ligita Espinosa, Robert Geier, Margaret Iding, Carly Pollack, and Heather Rhead.

A very special thank you to Nicole Gallicchio of NCDG Consulting who assisted in the final editing of the chapters of this book, especially at the very end when we needed to make our publication deadline.

Our university administration has also assisted us in peer-reviewing this education proposal and moving it forward to a single university competition at the National Science Foundation, the first of its kind to receive funding from the Partnerships for International Research and Education (PIRE) competition. Thank you, Douglas Gage, assistant vice president for research and graduate studies; Shobha Ramanand, research specialist; and Marjorie Wallace, director of research on teaching and learning.

There is also one very special person to whom we owe a huge amount of thanks—Richard Chester, the first project director of this study whose help in launching this work was immeasurable.

As in any study that takes place in schools, the individuals who make it happen are the students, teachers, and principals whose names confidentiality does not allow us to reveal—but to whom we owe the deepest gratitude.

Finally, we thank our funders, whose support has helped this project become a reality: the National Science Foundation and the Academy of Finland.

Team Members

U.S. TEAM

Christopher Klager

Tom Bielik

Deborah Peek-Brown

Israel Touitou

Kellie Finnie

FINLAND TEAM

Kalle Juuti

Janna Inkinen

Katja Upadyaya

Jukka Marjanen

Janica Vinni-Laakso

CRAFTING ENGAGEMENT FOR SCIENCE ENVIRONMENTS

CREATE for STEM Institute

"Crafting Engagement for Science Environments" (CESE) is the name of the study for which we were funded by National Science Foundation and which is discussed in this book. The CESE logo appears on the reports, curriculum, and assessments that we send out to researchers, policymakers, and teachers.

CREATE for STEM at MSU is the name of the institute at Michigan State University under which CESE is housed. The CREATE acronym stands for Collaborative Research in Education, Assessment, and Teaching Environments; STEM for science, technology, engineering, and mathematics. During the course of our research they provided legal advice, psychometricians, science experts, science teacher educators, and assistance with automated analysis of constructed response (AACR) for analytic purposes.

Why Learning Science Matters

The bell rings as first hour is about to begin. By twos and threes, students duck into Ms. Newman's classroom just in time to begin science class, while others surreptitiously trickle in after the bell. Ms. Newman takes attendance while students catch up on the latest gossip with each other. There is a quiet buzz in the classroom. Some students have their heads on their desks, enjoying the last few moments of rest before their day officially begins.

After a minute or two, once everyone has been accounted for, the daily drill begins as Ms. Newman calls out: "All right, everyone, let's get out the homework from last night." Most students dig through their backpacks, pulling out crumpled papers—with the exception of the valedictorian, who brings out a neatly folded, printed version. Some frantically try to fill in problems that they forgot to do the night before, with a few casually glancing at their neighbors' work for confirmation or to copy answers.

Ms. Newman places the stack of papers in her desk drawer and walks to the front of the room. "Okay, did anyone have questions about the problems on the homework? Let's do a couple together." Albert raises his hand. "Number two." Ms. Newman nods, then turns to the board. Over the next few minutes, she methodically goes through problem two, making sure to emphasize the procedural aspects of solving the problem so that students will be able to repeat its solution on next week's test. "Okay, any others? Let's try number six. That was a really tough one."

Again, Ms. Newman turns her back on the class and begins writing on the board, carefully explaining the steps. Some students follow along carefully in their notes, copying everything Ms. Newman writes on the blackboard. Others stare blankly—maybe paying attention, maybe not. "See, that was a hard one but it wasn't so bad, was it? Just like the first one, if you're careful." There are a few nods, as well as a few looks of *confusion and mild panic. "If there are no other questions, let's move on to something new. Let's talk about what's going on at the molecular level when a chemical reaction occurs." Most students appear to know what to do next as they turn to their notebooks, ready to dutifully copy down whatever Ms. Newman says, however cryptic, as she begins her lecture using the blackboard.*

Does this scenario seem familiar? It is one that has been replicated time and again over decades in secondary science classrooms: the same methods, the same content, and the same blank stares—or, in more extreme cases, a sense of dread and disengagement. Now contrast this with a morning in Ms. Xie's classroom.

Ms. Xie's students also duck through the door right as the bell rings, and catch up on last night's happenings with their classmates. Like Ms. Newman, Ms. Xie takes attendance—but then something different happens. She begins her chemistry class with a question: "Okay, let's get started. I'm having some trouble and I need your thoughts. I set a bottle of chemicals in here yesterday, but when I came back this morning, the bottle was still there but the contents had disappeared. Does anyone have an idea about where it went?"

"Someone probably stole it," says one student in the front. "Yeah, people take all kinds of stuff even if they don't know what it is," agrees a student in the back. "Maybe it spilled," chimes in someone else.

"Hmmm, those are interesting ideas, but I'm not 100 percent sure that's what it was. Look at this stuff," she says as she shows them a full bottle. "There's nothing special about this. It just looks like water. I can't imagine anyone would take it. And there was no puddle. It's just gone."

"Maybe it evaporated?" says a student. "Nah, when I have a cup of water by my bed it takes days to evaporate. There's no way a whole bottle could disappear overnight," another student interjects. Ms. Xie responds, "Now that's an idea. Let's try something." She pours a bit of the substance onto the counter, forming a small puddle. "Oh, I totally forgot about the homework. Can you all pass it up to the front?" Students dig through their bags to find the problems from the night before. Several students can be seen frantically filling in answers.

As the last few papers come to the front, Roger notices something. "Whoa, where did the puddle go? It's gone!" There is some laughter from the class and a few guffaws. "Well, I think we figured that one out."

Ms. Xie feigns relief. "I'm glad it wasn't stolen. But where did it go? I need it back!" More students laugh. "Okay, so if we know the liquid evaporated, let's try to figure out where it went. On your whiteboards, pair up and work together to write questions for what you think happened to the liquid." Students pull out their whiteboards and some begin to write immediately, while others plan with their partners about what to say. "Also, try to think of some other situations where substances appear or disappear and we don't know where they go. Write some questions you have about this phenomenon." After several minutes of working, Ms. Xie invites students to share their questions on a whiteboard, where they will become the driving questions for the unit on conservation of matter.[1]

Ms. Xie is not alone in her approach to science learning. Something new is happening in high-school science classrooms in the United States, Finland, and several other countries.[2] Over the past decade, a spate of international reports from the OECD, the European Union, and private European foundations have recommended major changes in science learning and instruction.[3] We are seeing a disruption of instruction, whereby teachers, administrators, parents, and policymakers are recognizing that traditional ways of teaching science are not enough to equip students with the rapidly changing content, technology, and life skills they will need for their futures.

This recent campaign for imperative science reform needs to be understood in light of prior U.S. science improvement efforts that, like those in Europe, have been characterized by "fits and starts," with change often followed by inactivity.[4] Nearly seventy years ago, during the Sputnik era, the United States found itself behind the Russians in space exploration. A quick response by policymakers, scientists, and educators launched a serious revamping of U.S. curricular content and instruction. The importance of science was embraced not only in schools, but also by the media: for instance, children's television programs like *Watch Mr. Wizard* were created to show how to do experiments, build rockets, and grow bacteria, while *My Weekly Reader—A Children's Newspaper,* designed for elementary students, envisioned flying cars, space stations, and domed metropolises protected from climate change.[5] A new science curriculum found its way into the schools, and student academic performance improved. Space science, computer technology, and engineering became promising career paths.

This emphasis on science exploration and learning was rather short-lived. By the 1970s and continuing through the first decade of this century, while the world of science learning was exploding, science instruction in schools remained somewhat resistant to these changes. There were some bright spots (for example, a growing emphasis on science in middle schools). For the most part, however, the world of scientific discovery, innovation, and change was not influencing school curricula. This gave rise to a fear that, without a major change, humans would be ill-equipped to harness the fruits of technology and meet the challenges presented by population growth, food scarcity, and the needs of a living planet.[6] These early and ongoing concerns were not exclusive to the United States—other countries also recognized that science learning needed to change. Finland, for example, despite its high student scores on international assessments of science knowledge, became increasingly worried about sustaining student performance and ensuring that high-quality science learning would translate into both scientific innovations and greater numbers of students pursuing careers in science.[7]

Over the past six years, U.S. educators, scientists, and policymakers have managed to turn science education on its head—an endeavor that, in contrast to other national efforts at reform, has been met with sizable acceptance.[8] This march toward a new set of standards is widely recognized as beginning in 2012, when the National Research Council (NRC), an independent division of the National Academies of Sciences, Engineering, and Medicine, produced a seminal report based on decades of research: *A Framework for K–12 Science Education.*[9] The report proposed a three-dimensional strategy that defined foundational knowledge and skills for K–12 science and engineering based on (1) scientific and engineering practices, (2) crosscutting concepts that unify the study of science and engineering through their common application across fields, and (3) core ideas in four disciplinary areas—physical sciences; life sciences; earth and space sciences; and engineering, technology, and applications of science.[10] The intent of taking such an approach to science education was to inspire students to make sense of phenomena and design solutions to problems.

Keeping the impetus for change in motion, the Next Generation Science Standards (NGSS)—an unprecedented set of standards created by scientists, educators, and policymakers—created a new vision for science education. The NGSS rejected earlier conceptions of learning science that focused primarily on the acquisition of scientific facts and a superficial understanding of scientific principles (such as the prevailing practice of centering instruction on memorizing principles and equations, with limited laboratory experiences and minimal opportunities for students to engage in "doing science to learn science").[11] This critique of outmoded instructional practices was coupled with a deep concern that science education was not tapping into the major advancements in science that were occurring daily, and which were rapidly accessible via technology. Reform was therefore a necessity, and the first strategic step was to formulate science standards that could help guide science educators and policymakers toward making significant changes in science education.

Focusing on in-depth and meaningful scientific learning, the NGSS described performance expectations that involved integrating the three-dimensional learning goals defined in the 2012 NRC report. Although the NGSS identified what students should know and be able to do, it did not prescribe a specific curriculum—the assumption was that curricula would need to be developed that enabled students to think and behave like scientists and engineers (that is, to explain phenomena and design solutions). By not proposing specific curricular content, the NGSS opened the opportunity for educators to reexamine, critique, and generate science experiences that would engage students and would be aligned with the standards.

Finland, like the United States, also brought together educators, scientists, and policymakers to redesign its science curriculum. The Finnish education system operates in collaboration with its ministry of education, municipalities, and schools, with teachers playing key roles.[12] Interested in advancing technology, en-trepreneurial activity, and environmental sustainability, the Finns began by devis-ing core aims and objectives for their elementary and lower secondary schools, then moved to create new science objectives for the Finnish National Core Curric-ulum for Upper Secondary School.[13] This national core curriculum highlights the need for students to actively acquire and apply science knowledge along with twenty-first century or generic competencies (attitudes, knowledge, and skills), with an emphasis on the use of technology in learning, both in and out of school.

Comparable to the NGSS, the Finnish curriculum and models of learning and instruction emphasize the design and use of science and engineering practices in order to support students in learning science, to prepare them for understanding the work of scientists, and to make science careers more interesting to them. This emphasis is echoed by the European Commission, which suggests that school science should better represent real scientific and engineering practices and cater more effectively to the needs and interests of young people.[14] Perhaps the greatest similarities between the two countries' standards and objectives are the consis-tency of their learning goals and the call for reform. But one important distinction remains (and will be discussed in greater detail later): in Finland, decisions about which scientific practices and curriculum content should be enacted in classrooms are made with the deep involvement of professional teachers, who have subject-area expertise and experience with empirical scientific research.[15]

Why Should We Care?

An additional explanation behind U.S. states' acceptance of the NGSS could be the release of science literacy results from the 2015 Programme for International Student Assessment (PISA) and Trends in International Mathematics and Science Study (TIMSS). Twelfth-grade science scores from the TIMSS have remained un-changed since 1995.[16] Similarly, PISA indicates that science learning has basically stagnated for secondary-school students in the United States: science scores have not improved since 2009, and U.S. students score only slightly above the average for all Organisation for Economic Co-operation and Development (OECD) countries.[17] This is especially troublesome given the importance of science knowledge for the development of the economy and for the acquisition of twenty-first-century life skills.

Finland's academic profile on PISA is strikingly different from that of the United States, but it faces a related problem. Among all participating OECD countries, Finland's science academic performance was ranked number one in 2003, 2006, and 2009. In 2012 and 2015, though its scores were still near the top (and considerably higher than those for the United States), they began to decrease slightly. It was not this slight decline that put Finland on notice about its country's science programs, however, but rather recent findings indicating that, among all OECD countries, the percentage of Finnish students who expect to work in science occupations at the age of thirty is the lowest. This weak interest in science, especially among females, is worrisome to the Finns—in particular, for the potential negative effects it may have on the supply of their science workforce, on support for scientific literacy, and on the overall health and well-being of Finnish society.

These PISA results cannot be taken lightly. The changing world of science and technology affects all our lives, as we interact with everything from computers and social media to the natural environment. Yet the funding streams for supporting science literacy are inadequate, and diminishing in some areas. The demand for scientists, engineers, and skilled technicians exceeds the supply—a problem exacerbated by a leaky pipeline that, for many students, begins early and continues throughout their academic career and into the labor market. But a deep, usable understanding of science is also critical for daily living in order to make decisions regarding the environment, health issues, and the safe and responsible production and consumption of goods.

If more students are going to appreciate the value and importance of science literacy and to consider science careers, they will need experience with scientific practices that will help them understand and interpret how the world functions. Critical to this process is stimulating students' imaginations and allowing them to work on thorny, purposeful problems with competing solutions. Both the United

States and Finland agree that the process of learning science needs to be dramatically revised, and that it should focus on developing usable knowledge to make sense of phenomena and find solutions to complex problems. One instructional method that aligns with the NGSS and new Finnish standards—and meets the challenges facing both countries—is project-based learning (PBL), an approach that encourages students to use their imagination by "figuring out" phenomena and solving problems using scientific ideas and practices.

What Can We Do about It?

Over the years, there have been multiple types of design-based science curriculum interventions.[18] What is different today is that both the U.S. and Finnish standards align with PBL, an intervention for science curricula conceived several decades ago that is experiencing something of a renaissance.[19] One of its earliest innovators is also one of our team leaders: Professor Joseph Krajcik, from Michigan State University, who served on the NGSS leadership committee, led the design team for the physical science disciplinary core ideas for the K–12 Framework, and has been directing our work with PBL.[20] He is joined by another of our team leaders, Professor Jari Lavonen at the University of Helsinki, who over the past thirty years has worked on national-level curriculum design and implementation in Finland.[21]

PBL aligns well with the three-dimensional learning strategies described in the K–12 Framework and NGSS, which detail an approach for making science learning more meaningful and authentic.[22] The first essential feature of a PBL unit is the driving question. The driving question is tied to a real-world situation that students find interesting and important, and students work together to find a solution to it over the course of the unit. The driving question allows students to raise other questions and increases their sense of wonderment about the world in which they live. It provides context for the tasks that students will do throughout the unit, and it gives these tasks continuity and coherence as the students build more sophisticated knowledge and understanding.

Another key feature of PBL is that the lesson-level learning goals are carefully tied to the learning performance expectations of the NGSS and the Finnish Core

Curriculum. The development of these lesson-level learning goals follows a specific and thorough process that begins with the selection of the national or state standard(s) that will be addressed by the unit. The core ideas of the standards are then unpacked—that is, broken down to their component concepts that are then individually explained and expanded. This ensures a thorough understanding of the core ideas before they are assembled into lesson-level learning goals.[23]

Much like the NGSS standards themselves, these lesson-level learning goals integrate a disciplinary core idea with a scientific practice and crosscutting concepts. The learning goals, however, are much more targeted, and serve as a guide for developing the driving question and tasks for the unit. Next there is an assessment process, which requires that students construct artifacts (for example, models or evidence-based explanations) that exemplify the intent of the driving question. Teachers are encouraged to develop daily assessments of the students' participation in these classroom activities, ensuring that students collaborate with one another to make sense of the data and form evidential claims. Finally, PBL has developed independent assessment tools that demonstrate the students' ability to articulate the driving question and their understanding of its answer, through a process that calls on the students to demonstrate scientific practices (such as building a model that has an evidential solution).

New studies of PBL suggest that it may be a viable solution for helping teachers to reconceptualize and organize their science instruction in ways that encourage students to use their imaginations and examine different points of view.[24] For example, in one PBL secondary-school physics unit, students explore the question "How can I design a vehicle to be safer by minimizing the force on a passenger during a collision?" They answer the question by analyzing data to support the claim that Newton's second law of motion describes the mathematical relationships between the net force on a macroscopic object, its mass, and its acceleration. In a chemistry unit, students develop and use models to explain that at the macroscopic scale, energy can be accounted for as a combination of the motion of particles (objects) and their relative positions.[25] As these examples demonstrate, PBL units require students to apply the "big ideas" of science and develop rigorous understandings of the content to make sense of phenomena.

We know from research in this area that students will not learn the content of their science courses, or scientific practices, unless they are actively engaged in constructing their understandings by working with and using ideas in real-world contexts. These ideas are explored more thoroughly in Part 2 of this book, which describes the transfer of these ideas into learning goals based on real-life situations. For students, thinking imaginatively about science and applying what they know, rather than memorizing facts and providing textbook answers, is challenging. It can even be a little uncomfortable. After all, individuals are being asked to solve problems that they would previously have been given the answers to—or have used well-known procedures to come up with a solution for—and now they must create a strategy for solving them. Yet the standards and their enactment through PBL emphasize that students seek new and innovative ways of "figuring things out," however challenging or uncomfortable. This idea of active knowledge construction suggests that understanding occurs when students can derive meaning for themselves based on their own experiences and, when they need to, put ideas together to make sense of data. But how can teachers offer these various situational experiences to students? And, more importantly, how will we know that students, by being involved, are deeply engaged in learning new ideas, participating in actual science practices, and able to incorporate different points of view?

How Are We Assessing Engagement?

As specified in the K–12 Framework and the NGSS, social and emotional factors are critical to the learning of science. Our other project team leaders have expertise in these areas. Professor Katariina Salmela-Aro, at the University of Helsinki, is a leading authority on the psychological well-being of adolescents and adults.[26] Professor Barbara Schneider, at Michigan State University, is a key figure in evaluation design and an expert in conceptualizing and measuring adolescent development.[27] These two scholars have both used innovative techniques to assess social and emotional learning and are bridging science learning with the affective aspects of our work.

One of the most important recommendations of the NGSS is for students to become more engaged in science, to learn why science is important to them, their futures, and our society. But what does it mean to be engaged? What is engagement? And how can we know when it is occurring? Psychologists have multiple meanings for engagement and employ various measures to determine when students are actively involved in ways that impact their performance.[28]

We define engagement as having three critical properties: interest, skill, and challenge. Interest is a predisposition for a specific phenomenon, such as wanting to know what causes some materials to collapse upon impact, or why beehives seem to be disappearing. Skill is the prerequisite knowledge needed to master new learning opportunities—for example, knowing the fundamentals of Newton's laws of motion and being able to apply them when creating a new model. And challenge is a course of action in which results are not entirely predictable but are testable, like when planning and conducting an experiment to explain a phenomenon.

In contrast to others who have conceptualized engagement as a general trend, we limit our measure of engagement to precise moments occurring within specific times and varying in intensity: this allows us to, for example, compare discrete levels of student engagement between listening to the teacher present ideas versus building a model. We conceptualize highly engaging times as optimal learning moments (OLMs): situationally specific instances when an individual is so deeply engrossed in a task that time feels as if it is flying by. This approach is similar to how Mihalyi Csikszentmihalyi described the state of "flow" as one of being completely immersed in an activity.[29] For our purposes, however, we are restricting this definition to moments in learning situations that elevate students' social, emotional, and cognitive learning.

We believe that carefully crafted instructional situations can enhance engagement and that, when this occurs, learning deepens.[30] It is an empirical question, however, as to how frequently students are engaged and what types of instruction can enhance engagement. PBL's alignment with both the NGSS and Finnish standards has enabled us to begin evaluating, in both countries, whether it has a more positive impact on engagement and depth of learning than more traditional science instruction.

Beginning with the 2015–2016 school year, we have been testing the impact of

PBL on student engagement in several chemistry and physics classes in the United States and Finland, using a variety of methods to obtain measures of students' social, emotional, and cognitive learning when doing PBL and comparing them to measures of when students are involved in more traditional science lessons. We are also in the process of developing new designs for summative assessments related to cognition. To maximize diversity among our U.S. student populations, we used the Michigan Department of Education's administrative state data to identify our schools. At least 30 percent of the sampled school population in the United States is minority and low-income. While Finland has a more homogenous population, the Finnish schools in our sample have experienced a slight increase in immigrant students, and reflect a somewhat wider range of family social and economic resources. We use multiple instruments for testing our intervention, including the experience sampling method (ESM), a time-diary randomized survey-sampling technique administrated via smartphones. The ESM provides a mechanism for learning what students are actually doing in the moment and how they feel about it. The development of this ESM is elaborated in Chapter 4.

In addition to the ESM, which is given to both students and teachers, we obtain information from student and teacher surveys that examine basic demographic information and aspirations, career knowledge, and interest in STEM. Specific questions ask students to describe their teachers: Are they caring? Do they treat students with respect? Do they keep the class busy? Can they explain difficult topics clearly? Do they help students correct mistakes? Do they feel burned out or stressed in their job? Teachers are also interviewed, and they share lesson logs and plans along with other classroom materials, including the products that students develop in both business-as-usual and PBL lessons.

Along with the assessment tools, our team has been developing pre- and post-PBL assessment tasks to be part of the PBL lessons. These assessment tasks require students to make use of three-dimensional learning to make sense of phenomena or solve a problem. The assessment tasks relate to the broad performance expectations of the NGSS but are given in a different context than the students were exposed to in their units. These assessment items have been critiqued by science educators and researchers, as well as chemists and physicists. Cognitive laboratories with teachers and students have also been set up to determine whether the pre- and post-assessment questions match the tasks that students are being asked to undertake.[31]

Our international team has worked jointly on the development of identical instruments and technologies, and has translated many of the English documents into Finnish, making sure that the meanings of words are consistent across both countries. The intervention units were created in collaboration with teachers and science-teacher professionals in both the United States and Finland. This development process took place over several years, beginning with two full days of professional development, teacher exchanges between the United States and Finland, virtual collaborative meetings, and regular updates on activities. Throughout the project there has been consistent feedback on and modification of the PBL units. Fidelity of implementation was conducted on a selective randomized video schedule, in which coded instruction was triangulated with student and teacher ESM responses and interview data.

To test the effectiveness of PBL, we use a single-case design that requires the repeated application of the intervention between specified times.[32] The data are collected from the students during business-as-usual instructional periods and then during the PBL intervention units. This pattern is replicated several weeks later for the same amount of time, in both conditions. While single-case designs are appropriate for this stage of our study, in the next phase of our work we will be scaling to multiple states, reaching thousands of students and conducting a cluster-randomized trial.

Results so Far

One might ask why we are releasing these results before reaching the final stages of a larger effectiveness study. The answer: our results to date have been so promising that it seems almost irresponsible not to adopt an improvement science framework (as initiated by Anthony Bryk, president of the Carnegie Foundation for Teaching).[33] With a project of this complexity, which involves two countries, the data we have collected, analyzed, and published findings from show a pattern of PBL's consistent positive effects on students' social, emotional, and cognitive learning.[34] Not only are the students experiencing positive effects from PBL, but their teachers also report that PBL has been transformative. The experience of these teachers—and the positive effects it is having on their students—is a story that needs to be told.

Our book is written not only for academics but also for professionals interested in aligning their science learning with PBL. Additionally, it is for parents, so they can learn what types of science-related questions and activities are likely to interest and engage their adolescents. We have spent considerable resources to develop our open-source units, which we are eager to share: we have already started the process of giving our science materials to the OECD as part of their new creativity project.[35] We are also finding that our project is often oversubscribed—that is, more teachers want to participate than we have the resources to test in our empirical work. In response, we have agreed to distribute our materials to teachers who want them and would like to become part of this initiative in subsequent years.

We have titled our book *Learning Science* because the NGSS and Finnish aims were both based on the emerging scholarship on how to create classroom environments where people learn effectively. Cognitive science research on the neurological processes of learning are revealing the importance of individual, cultural, and technological influences on the brain as it adapts to the structure of learning environments. We are committed to using these insights to formulate our intervention and to articulate what students should be expected to accomplish in science and engineering environments. By incorporating recommendations from the learning sciences, we have created plans for learner-centered classroom experiences that are well organized, underscore doing with understanding, use formative assessments to give students feedback, uphold norms of cooperation and problem solving, and help teachers and students formulate and answer meaningful questions.[36] PBL is designed to operationalize these ideas by incorporating three-dimensional learning to make sense of phenomena and solve problems using science-based strategies. Our goal in writing this book is to give educators and parents the tools they need to advance science learning, regardless of future policy trends or constraints—and to show policymakers, in turn, how curriculum developers, educators, and other practicing professionals can work together to transform science education with rigorous and meaningful evidence-based work.

PART I

Transforming the High-School Science Experience

1. Creating Science Activities That Engage and Inspire

It is a typical Midwestern winter Saturday morning and Kristie, a high-school physics teacher, opens the door to the school district central office building. Outside it is cold and snowing, but the weather is not going to stop her from learning how to get her students more engaged in science. She looks across the room and sees a colleague who teaches physics at a different school. "Hi, Tom, so you're here, too?" Tom replies, "Yeah, I wouldn't want to miss this—I've always been intrigued with project-based learning and this seemed like a great way to learn more about it and find out how I could use it in my classroom." Kristie responds, "Me too, especially now that our school district wants to have all of our science teachers adopt the new science standards. From what I've read in The Science Teacher, *project-based learning can help us link these standards to our curriculum and instruction."*

After introductions and a review of NGSS and PBL principles, the teachers form teams of twos and threes and quickly become actively involved in a PBL physics lesson, developing scientific models using tools on the computer. Kristie and Tom are huddled over the computer, trying to explain what happens when changing the velocity and mass of cars in a collision. Kristie says, "This is really good. All our kids drive, but unfortunately not always safely. This is actually really relevant to them." Tom nods, saying, "I didn't even know that something like this computer modeling program existed! I am really psyched to do this with my kids—they're really going to have to figure out what happens when you change the velocity of a car. It'll be tough, but I bet they're really going to enjoy doing this."[1]

Kristie laughs, "Hey, I like it! Let's change the angle of the ramp and see what hap-

pens! Sort of like coming down that steep stretch before the curve on Abbott Road. Tom,
what do you think about changing the angle?"

Although this type of activity is commonly called "professional development,"
we have adopted the term "professional learning communities" because it aligns
more closely with PBL and its emphasis on collaboration and shared understandings of phenomena by both teachers and students. Professional learning is essential for enacting and sustaining PBL because it provides teachers with "hands on"
experiences to hone their knowledge and skills and support student engagement.

PBL has had an interesting history both in the United States and in Europe.
Some of the earliest citations regarding the project method can be found in essays
written by William Heard Kilpatrick, which have been traced to publications and
speeches prior to 1917.[2] A student of John Dewey, Kilpatrick was one of the first to
encourage the use of purposeful project activities in a classroom—when carried
out in accordance with conventional established knowledge and skills. Dewey's
critique of Kilpatrick's proposition was that it embraced too much focus on "un-
restricted student choice." While choice was viewed as important, Dewey argued
that it was not unconditional. Instead Dewey called attention to the "act of thinking," a process whereby students encounter a problem, plan a solution, work on
it, and reflect on the results.[3] The Dewey framework, with its emphasis on learning
through inquiry-driven projects, is the one typically referenced when tracing the
roots of PBL.[4] While Kilpatrick was immensely popular (there are reports of his
work being circulated to over fifty thousand individuals), his message of students'
untethered freedom and project learning as a solution for obtaining knowledge and
attitudes failed to achieve enduring legitimacy. Some have even suggested that it
formed a basis for the discrediting of education science.[5]

European support for many of the ideas found in PBL can also be traced to
John Dewey, although the European model focuses on conceptual terms commonly
identified as inquiry-based and/or problem-based learning. Researchers in Europe
have, for the most part, identified two types of pedagogical approaches: deductive,
whereby the teacher transmits knowledge to the students; and inductive, whereby
students are given more opportunities to observe and experiment, with both stu-

dents and teachers more involved in knowledge construction. The inductive approach has been termed inquiry-based science education (IBSE). Adopted from mathematics, problem-based learning describes an environment where problems drive the learning—and students do not seek a single correct answer but rather interpret problems, gather information, identify solutions, evaluate options, and present conclusions. Problem-based learning and project-based learning are not synonymous, nor are we implying that they are—but they do share some pedagogical approaches that are clearly distinct from more traditional deductive practices.[6]

New European proposals for science education reform have promoted a perspective comparable to that articulated in PBL and the NGSS.[7] European recommendations underscore the importance of engaging students in science experiences that involve extended investigative work and "hands on" experimentation, both of which offer coherence and relevance to the task of exploring the implications, application, and use of scientific knowledge.[8] Most recently, PBL pedagogical practices and professional learning experiences have received considerable recognition and acceptance across multiple countries, specifically within the Finnish education system.[9]

What Is PBL and Why Does It Lead to Engagement?

PBL, which combines Dewey's ideas with insights from learning science research, gained major attention with the design-based movement of the 1990s, which advocated experiential approaches to education.[10] The design components for our PBL intervention have their roots in an article from 1991, in which Phyllis Blumenfeld and her colleagues compellingly linked the idea of supporting learning by having students engage in "long-term problem-focused, meaningful units of instruction that integrate concepts from a number of disciplines or fields of study" and explored how technology can be used to sustain motivation and thinking.[11] Indeed, many of the ideas promoted by Blumenfeld and her colleagues in their conception of PBL twenty-five years ago are present in the *Framework for K–12 Science Education* and the NGSS framework of today.

Although a number of PBL studies were conducted between 1990 and 2000, as John Thomas noted in his 2000 review, they seemed to lack a uniform vision of what PBL actually entails. He concluded that while PBL appeared promising, results were not robust enough to prove that its effects improved student outcomes (such as science achievement). It was suggested that PBL's inconclusive findings were primarily the consequence of a shortage of strong design models and valid, reliable measures.[12] What PBL is, and is not, has prevented it from achieving a commonly agreed-on definition.[13] But given the importance placed by the *Framework for K–12 Science Education* and the NGSS on deeper learning through experiential approaches to science learning, teachers' and policymakers' understanding of what PBL is has become more relevant and compelling.

The *Framework for K–12 Science Education,* as well as stakeholders' increasing concerns about decades of lackluster performance by U.S. students in science, a general lack of scientific literacy and technological knowledge, and labor shortages in some STEM careers, have energized an effort to create a new set of national standards for science learning.[14] As explained in the opening chapter, it is not surprising that the public and academy were supportive of the NGSS, because it calls for deeper learning acquired through meaningful experiential activities. At the heart of the NGSS is the call for three-dimensional learning, which employs core scientific ideas, crosscutting concepts, and scientific practices to guide learners as they make sense of phenomena or solve problems.[15] The standards, however, do not specify how this should be accomplished (for example, by using a specific curriculum or employing certain assessments). Our intervention takes the next step by articulating the design-based principles that link our PBL curricular units with the NGSS and the Finnish Core Curriculum, including their learning goals, activities, and assessments.

In Finland, there also has been concern among stakeholders regarding students' science interest and careers, and the Finns too have recently revised their science curriculum to encompass many of the same principles as do the *Framework for K–12 Science Education* and the call for more experiential activities in science learning.[16] The new Finnish framework emphasizes the importance of student engagement in science learning—an engagement supported through a focus on core scientific knowledge and practices, and through active participation in project-based learning.[17] The acknowledgment of problems with science instruction and the emergence of new standards in both countries forged a basis for our collaboration, spurring the mutual exchange of ideas, curricular practices, and assessments.

Our collaboration began fortuitously at an international meeting funded by the National Science Foundation and the Academy of Finland in 2014—the topic of which was enhancing engagement in science learning. Discussions between scholars invited from Finland and the United States reached a common ground quite quickly, with both countries' representatives recognizing the problems of enhancing science interest, especially among females, and sustaining science careers through college and the labor market. Building on cross-national networks, a diverse interdisciplinary Finnish-U.S. team of researchers experienced with science education research, social and emotional development, and teacher education came together with the shared goal of conducting a study to test a system to advance science learning for teachers and students via enhanced engagement.

Our international team recognized the importance of redesigning curricular units, assessment practices, and teacher professional learning in ways that would reflect the new standards and could be evaluated in a multistage process. Although there are considerable differences in the ecological landscapes and education systems of the two countries, our team developed a study whereby each country could employ a similar research approach that included an equivalent treatment, that is, PBL and an evaluation plan. Over the course of several months, the team found there was considerable crossover with regard to research methods; expertise in instrumentation for measuring academic, social, and emotional learning; and assessment strategies. Upon easily reaching a consensus, the team decided on the subject matter that would be the focus of the research design, the age of the students in the study, and what types of professional learning experiences would be created and used with participating teachers.[18] Both the Finnish and U.S. researchers decided to work together to create units that would emphasize the kinds of three-dimensional learning described in the NGSS and would complement the Finnish Core Curriculum. It was not expected that the PBL practice would be identical across countries—rather, the design would need to accommodate each country's school-based formal structural and informal cultural contexts.

After a year-long process of working together on the design of the study, including the development of the PBL units—for which Professor Krajcik took the leadership role—the team was ready to begin a field test. In both Finland and the United States, the teams reached out to teachers to participate in the intervention. The Finnish teachers, who frequently interact with university researchers, were genuinely interested to learn new instructional approaches for enhancing science interest among their students.[19] And the U.S. teachers were committed to finding approaches to science learning that would match the new national and state standards that were being implemented in their districts.

The Design Principles of PBL

In keeping with the principles of design-based studies—which underscore the importance of defining claims about teaching and learning as well as the relationships between theory, activities, and artifacts/products—our PBL U.S.-Finnish intervention is an instructional approach with lessons designed to engage students in relevant and meaningful learning experiences that provide solutions to real-world problems.[20] The design principles of our PBL intervention focus on six factors.[21]

1. *Meeting important learning goals.* When designing the units, we began by considering the NGSS or Finnish Core Curriculum performance expectations that would be their focus. The identification of these performance expectations is a critical step, because they link the driving questions to scientific practices and core ideas within the discipline: for example, students are asked to build a model to explain why water evaporates more slowly than alcohol, rather than just learning the facts surrounding the motion of particles. In this way, performance expectations shift from being simply content standards with little contextual meaning to being learning goals supported by explicit scientific practices. These lesson-level performance goals drive the tasks that occur in each lesson and build across the unit.

2. *Constructing a meaningful driving question.* By pursuing a solution to a problem or an explanation of intriguing phenomena, students exer-

cise the scientific practice of asking questions to identify the information needed to complete the task. Developing a meaningful driving question is the most fundamental component of PBL, because the question constitutes the anchoring phenomenon or problem that students will attempt to "figure out" throughout the unit. The driving question must be relevant, explored in-depth, lead to the asking of additional questions, and be threaded through the various lessons of the unit. In other words, the driving question should provide coherence across multiple topics and scientific practices.

3. *Providing opportunities for learners to explore the phenomenon using scientific practices.* To answer the driving question (and smaller lesson-level ones), students draw on scientific practices such as planning and carrying out investigations, analyzing and interpreting data, and constructing explanations and designing solutions. Among these scientific practices, modeling and evidence-based explanations have been shown to be valuable for giving students opportunities to explain and predict phenomena encountered in PBL.[22]

4. *Creating collaborative activities that will help students find solutions to the driving question.* Students' ability to learn science is enhanced by collaboration that allows for rich discussions and the exchange of ideas to build knowledge—just as the work of actual scientists and engineers involves considering a variety of perspectives and testing claims to develop evidence-based solutions. In PBL classrooms, students collaborate to make sense of data and information they have gathered in order to design answers to the driving question. These actions align with various scientific practices including collecting and analyzing data; constructing models; arguing from evidence; developing explanations; asking questions; and obtaining, evaluating, and communicating information.

5. *Integrating learning tools to make sense of evidence in unique ways.* Incorporating technologies can help teachers foster inquiry and student learning. Learning tools include, for instance, video simulations of interactions between forces in physics; the motion of particles in chemistry; and constructing and testing system-based models with computer software. These and other tools can increase students' use of technologies that they are likely to encounter in our increasingly complex scientific world.

6. *Using tangible artifacts or assessment tasks to address the three dimensions of learning and capture students' emerging understanding.* Artifacts and other assessments take multiple forms, and their purpose is to immerse students in the world of scientists and engineers. These culminating products allow students to demonstrate the knowledge they have ac-

quired, the skills they have developed, and the scientific practices that they have learned to use to make sense of the phenomena and to solve problems—not just in science class but in the wider world of big ideas.

Why PBL for Chemistry and Physics?

We chose to design six units (three in chemistry and three in physics), which we refer to as the Crafting Engagement in Science Environments (CESE) curriculum, to improve science achievement, engagement, and other social and emotional learning experiences. We selected chemistry and physics for a number of reasons. To begin with, many states have instituted new curricular changes to help improve science and mathematics achievement and are now requiring students to take more advanced level courses in these subjects.[23] These subjects are thus often gatekeeper courses for U.S. postsecondary schools, that is, students must typically take at least one of these subjects in high school to be considered for admission to a competitive college.[24]

Although the Finnish secondary education system, detailed later, is markedly different from the U.S. system, researchers and policymakers share some of the same concerns regarding the social and emotional factors surrounding the teaching and learning of chemistry and physics. At the secondary level, the Finnish system consists of two parts: public comprehensive lower secondary school for grades seven through nine, and upper secondary school for grades ten through twelve. Science in the lower secondary system is subdivided into physics, chemistry, geography, biology, and (more recently) health education.[25] Broad core curricular guidelines are published for each of these two school systems, but the local education providers (that is, the municipalities) are typically responsible for curriculum design. There are few private schools, so the overwhelming majority of students attend common comprehensive schools. After completing ninth grade, students must choose whether they will attend upper secondary school or vocational school; this choice is usually split quite evenly, with half of the school population matriculating to upper secondary school, and the other half attending vocational school.[26] In upper secondary school, one course each in physics, chemistry, and biology is required for all students, and students can select more courses based on interest. One course consists of thirty-six lesson hours. This is where we see some commonalities between the United States and Finland: Finnish students often express the same disinterest and personal dissatisfaction about these required courses that many U.S. students have concerning physics and chemistry.[27]

According to recent research by our team members Kalle Juuti and Professor Jari Lavonen, although Finnish teachers have considerable latitude in designing their own classroom activities and students are not required to take a national exam at the end of the year, students report that experimentation in science appears to be teacher-led—where students follow steps detailed by the teacher instead of trying to design experiments of their own. Students in upper secondary school also report feeling that science, especially physics, is difficult, uninteresting, and unpleasant—but they view it as important.[28] Juuti and Lavonen conclude that the ideas underscored in the *Framework for K–12 Science Education* have real value for Finnish science classrooms. They stress the value of the framework for encouraging students to carry out investigations by themselves, set goals, and take responsibility. Rather than placing value on the instrumental side of physics, the authors conclude that if Finnish students become more engaged in science and find it enjoyable, they are more likely to perceive it as important to their futures.[29]

This problem of interest is critical to our present work, because research shows a relationship between secondary-school students' interest in science and whether they pursue science in postsecondary school and as a career.[30] There are real workforce concerns about technological training in both countries, especially regarding opportunities for individuals to develop and exercise their imagination and creativity in science fields, because these abilities help lead to innovation.[31] The standards in both countries stress these learning opportunities as valuable, but the topic is rarely addressed in current science teaching. Both Finland and the United States struggle to encourage workers, especially female workers, to pursue careers in science. We argue that PBL has the potential to increase interest in science—as well as creativity, literacy, and achievement in scientific disciplines—among groups that have historically been underrepresented in chemistry and physics, including women and minorities.

A Snapshot of Our PBL Units

Over the course of our pilot year, 2015–2016, four curricular units were developed and enacted, two in chemistry and two in physics. These were subsequently revised, and two additional subject-specific units were created in 2016–2017. Each unit was developed to last two to three weeks. In Finland, high-school courses are shorter—with each one lasting only several weeks and covering a narrower range of topics than the semester- or year-long courses typical in the United States—but individual lesson periods tend to be longer. As a result, the actual class time spent on the PBL units was similar in both countries, even though on a day-to-day basis the units were slightly different (to accommodate differences in course structure).

In conjunction with our principal investigators and other team members, Deborah Peek-Brown—a curriculum specialist—along with postdoctoral fellows Tom Bielik and Israel Touitou, led the curricular efforts in the United States (Peek-Brown and Bielik primarily led on curricular construction, Touitou on assessment). Juuti, in conjunction with several master's level teachers, led the effort in Finland with regard to both curriculum and assessment, and all materials were shared between the countries.[32] We relied on a number of team members to undertake the translation process so that we could review each other's work. The following brief descriptions are designed to give the reader a sense of how each of the six units cohere to the NGSS and Finnish standards. It is important to underscore that we agreed to use in both countries the PBL framework, which specifies performance expectations (Finland doesn't use the word "expectations," but the definition of what is meant by the ideas is very similar), driving questions for the units, the tasks that students do in the units, and the artifacts that capture students' emerging understandings.

The unit materials we have developed are more complex than what is described in the following snapshot—for example, we unpack the performance expectations, develop lesson-level performances for each of the daily lessons, and construct a storyline to assist the teachers in following the logic and coherence of student understanding that builds throughout the unit.[33] There are several important points that may not be obvious from these descriptions:

- The driving question is the basis for the daily lesson plans and it is threaded through the entire unit. The question is specifically created to be meaningful and relevant to the students' lives.

- In each of the units, students are actively engaged in building models, constructing evidence-based explanations, designing and setting up experiments, and collecting and analyzing observations to support evidential claims.

- Students create artifacts but our assessment process goes one step further: we created assessment tasks that appear on a pretest at the beginning of a unit, a post-test at the end of a unit, and a summative year-end assessment. These assessment tasks require students to make use of the three dimensions of science learning to explain challenging phenomena or solve perplexing problems.

Our first chemistry unit is evaporation. In the United States, this unit was designed to meet national standards, specifically NGSS PE HS-PS1–3: "Plan and conduct an investigation to gather evidence to compare the structure of substances at bulk scale to infer the strength of electric forces between particles"; and NGSS PE HS-PS3–2: "Develop and use models to illustrate that energy at the macroscopic scale can be accounted for as a combination of energy associated with the motion of particles (objects) and energy associated with the relative positions of particles (objects)." Finland developed its own goal for this unit.[34]

For this phenomenon to be relevant to students, there needs to be a driving question—one that is comparable in both countries. Here the driving question was "When I am sitting by a pool, why do I feel colder when I am wet than when I am dry?" With respect to unit phenomena, students use classroom experiments and models to figure out how evaporative cooling occurs and build understandings of interactions at the particle level and how they pertain to matter's macro-level structure and properties. Finally, to learn and assess how students make sense of phenomena, throughout the unit students construct models and explanations of the process of evaporative cooling, connecting the energy changes to changes in the structure of matter in the system.[35]

The next unit is conservation of matter, and its focus is NGSS PE HS-PS1–7: "Use mathematical representations to support the claim that atoms, and therefore mass, are conserved during a chemical reaction." The driving question for this unit is "Can I make substances appear or disappear?" Students start the unit by developing an initial model to explain why no ash is left after a piece of flash paper is ignited. Students design experiments to develop new ideas of what happens to matter at the atomic level. Using this information, students revise their models using an online modeling program to "figure out" why substances seem to appear or disappear. Students construct a final model (an artifact) that presents an answer to the driving question and an explanation supporting that model.

The final chemistry unit is about the periodic table of elements, based on NGSS PE HS-PS1–1: "Use the periodic table as a model to predict the relative properties of elements based on the patterns of electrons in the outermost energy level of atoms"; and NGSS PE HS-PS1–2: "Construct and revise an explanation for the outcome of a simple chemical reaction based on the outermost electron states of atoms, trends in the periodic table, and knowledge of the patterns of chemical properties." This unit's driving question is "Why is table salt safe to eat, but the substances that form it are explosive or toxic when separated?" Students work together and use observational data, classroom demonstrations, and group research to explain patterns in the reactivity of elements. Students figure out trends in the periodic table, and build an understanding of the relationship between the element's location in the periodic table and its properties and reactivity. For the final task, students construct a model that describes the relationship among atomic structure, electronegativity, and ionization energy and the reactivity of elements.

The first physics unit is Forces and Motion, and it targets NGSS PE HS-PS2–1: "Analyze data to support the claim that Newton's second law of motion describes the mathematical relationship among the net force on a macroscopic object, its mass, and its acceleration"; and NGSS PE HS-PS2–3: "Apply scientific and engineering ideas to design, evaluate, and refine a device that minimizes the force on a macroscopic object during a collision." The driving question is "How can I design a vehicle to be safer for a passenger during a collision?" Here, student activities include working collaboratively to investigate and develop computer models to explain collisions and to design a vehicle to be safer for passengers. To achieve this goal, students base their work on questions they generate, and go through the process of "figuring out" how to design a safer moving vehicle and build an understanding of the relationships among force, mass and acceleration (that is, Newton's second law), momentum, and impulse. Students use this knowledge in combination with engineering practices to develop their best design and then, using a given set of materials, they build and test a cart that minimizes the force on a passenger during a collision.

The second physics unit is magnetic fields, which focuses on NGSS PE HS-PS3–5: "Develop and use a model of two objects interacting through electric or magnetic fields to illustrate the forces between objects and the changes in energy of the objects due to the interaction"; and NGSS PE HS-PS3–2: "Develop and use models to illustrate that energy at the macroscopic scale can be accounted for as a combination of energy associated with the motion of particles (objects) and energy associated with the relative positions of particles/objects." The driving question here is "What makes a Maglev train float?" Students use the phenomenon of magnetic levitation to explore the characteristics of magnetic fields. In trying to answer the driving question, students "figure out" how magnetic levitating trains work and build understandings of the relationship among magnetic fields, force, and energy using classroom experiments and computer models. At the end of the unit, students construct a device that floats, using magnets, and explain the physics ideas behind it.

The final physics unit is electric motors, which is designed around three NGSS performance expectations: NGSS PE HS-PS3–1: "Create a computational model to calculate the change in the energy of one component in a system when the change in energy of the other components and energy flows in and out of the system are known"; NGSS PE HS-PS2–5: "Plan and conduct an investigation to provide evidence that an electric current can produce a magnetic field and that a changing magnetic field can produce an electric current"; and NGSS PE HS-PS3–3: "Design, build, and refine a device that works within given constraints to convert one form of energy into another form of energy." The driving question is "How can I make the most efficient electric motor?" Using electric motors as an anchoring phenomenon, students explore their electric and magnetic components through inquiry-based investigation. Using class experiments and computer models, students figure out how to build a more efficient electric motor and in so doing form an understanding of the relationship between electric current and magnetic fields. Finally, students use the concepts of systems and energy transfer to construct an electric motor that will be as efficient as possible given its material constraints.

The goal of the assessments is for students to show how they go about "figuring out a specific phenomenon," solving problems, and using evidence to support their claims and solutions, so that we can measure the effect of PBL on how well they have learned scientific ideas and mastered scientific practices. All of our assessment items are three-dimensional, requiring in-depth responses and, in many instances, the actual development of models. As with all of our procedures, our Finnish and U.S. teams work collaboratively on assessment items—in Finland, this process is being led by Lavonen and Juuti.[36] The construction of our assessment tools and processes is based on procedures developed by Professor Krajcik and colleagues.[37] The design of specific assessment tasks for our units is explained in a recent paper led by postdoctoral fellow Israel Touitou.[38]

The process we used to build and validate our assessment items included expert reviews by team members, scientists, and science educators. In addition, we conducted cognitive interviews with teachers and students to revise and refine our assessment items. Similar cognitive interviews were also conducted in Finland to learn if the questions were interpretable and yielded comparable feedback, especially with regard to how clear or difficult the questions seemed. All of our assessments allow students to write full-paragraph descriptions and draw models. Items are then scored using a rubric that assesses the students' knowledge of the disciplinary core idea, the process they use to obtain their evidence, and the subsequent result.

The Role of Professional Learning Communities

Presently, most teachers in both countries are unfamiliar with the process of developing PBL units. As part of our work, we have strategically orchestrated a set of professional learning workshops in the United States and Finland to support teachers in understanding the components of three-dimensional learning and its relationship to PBL. Similar to the student PBL units, these professional learning experiences are designed to provide opportunities for teachers to learn the purpose of and techniques for specifying learning goals, identifying meaningful problems, presenting challenges, employing scientific practices, and underscoring core ideas within the discipline. In addition to emphasizing the value of PBL for science learning, our goal is to actively support teachers in discussion and reflection, so they can learn from each other through collaboration and sharing experiences.[39]

The process of recruiting schools and teachers was relatively straightforward. Much to our surprise, we had many more U.S. schools and teachers than we expected who were willing to learn about PBL, use it in their classrooms, and measure its impact on student social, emotional, and cognitive learning. The process of recruiting teachers in Finland was equally smooth. In the United States, we contacted secondary schools located within a reasonable driving distance from our university. A similar approach was taken in Finland, where the schools were in close proximity to the university. High schools in both countries were selected to include variations in characteristics like student race and ethnicity, socioeconomic status, school location, and size. For while Finland has a relatively homogenous population, the country's recent increase in immigration has resulted in increased variation across its schools in terms of student social and economic resources.

Throughout the past two years, each country has had a series of professional learning meetings and eight international workshops where teachers shared their experiences working with the PBL units. Several of the teachers visited each other's classrooms in their respective countries. In the first professional learning meeting, Professor Krajcik and Deborah Peek-Brown provided meaningful context for the project, describing the history and contents of the *Framework for K–12 Science Education* and the NGSS and paying particular attention to the learning performances. This process was fundamental, because the performance expectations expressed in the NGSS propelled the development of learning goals for the PBL units.

The preparation of teacher guides and required materials lists was also initiated during these workshops. This process continued throughout the year through online meetings and frequent communication among participants, who provided feedback to the unit's lead teacher. This important feedback, based on teachers' experience and ideas, led to the refinement of the curricular materials to better meet the needs of the teachers and their students.

What Are We Learning?

A recent paper, led by Tom Bielik, details the professional learning of the U.S. teachers and their reactions to it.[40] Bielik conducted in-depth interviews with a select group of lead teachers and several others who participated in such ongoing professional learning experiences. Although experiencing several challenges around learning a new method of instruction, the teachers were optimistic that PBL was helping students to be as "open-ended as possible," and to avoid searching for clear-cut answers. Bielik's findings, though not to be taken as definitive, foreshadowed other empirical results we found using extensive teacher and student data (see Chapters 4 and 5).

Data were also collected from the Finnish teachers who used PBL with their students.[41] A team led by Professor Lavonen surveyed teachers, collected lesson plans, and conducted observations in teachers' classrooms. Finnish teachers talked about the positive influences of PBL, especially students' implementation of scientific practices. They reported feeling that PBL had led to a renewal of their teaching methods and had succeeded in making science more engaging for their students. The teachers were positive about their experiences and those of their students even as they described the challenges of taking on a new type of instruction and expressed a desire for continued support from researchers.

One reason that Finland's teachers are so willing to participate in professional learning activities and experiment with PBL in their classrooms can be traced to their educational backgrounds and the respect they receive from society. Finnish science teachers have a master's level education, majoring in the subject that they are teaching and minoring in educational sciences. To complete their degree, they have to prepare a thesis in both subjects. They are expected to develop their own curriculum and assessment procedures in accordance with broad policies set by the Ministry of Education and local municipalities. Teachers are regarded as experts in curriculum development, teaching, and assessment across all grade levels. A culture of trust by education authorities and national education policymakers sup-ports the view that teachers, together with principals and parents, are the best possible experts for determining the ideal educational content and practices for children and youth.[42]

Although we have taken the role of professional learning seriously, it is always somewhat unclear whether the teachers are implementing PBL with fidelity to its design principles, performance goals, and unit lessons. To measure this, we have chosen to use the experience sampling method (ESM), which gives us a "window" into each classroom and includes both teacher and student perspectives. We also collect lesson plans from the teachers and videos of classrooms to assess whether the intervention is consistent with the goals of PBL. And after a unit is completed, we interview the teachers about the challenges and successes encountered during the lessons. The result is a portrait of three-dimensional learning in action, with classrooms acting as dynamic ecosystems that reflect individual teachers' peda-gogical styles, school practices and routines, and the embedded historical and cul-tural behaviors and customs of each country.

2. Project-Based Learning in U.S. Physics Classrooms

Mr. Cook looks over his lesson plans, knowing that it will not be a normal day of reviewing physics homework and showing a demonstration to illustrate a concept. Today he will begin teaching project-based learning lessons that he has planned collaboratively with other physics teachers and university researchers during professional learning sessions. Over the years, he has been reading and hearing about PBL—now he will get a chance to try it out for himself and his students.

He reviews the lesson plans once more, to make sure that he has everything he needs for the first lesson. Slightly anxious, he is still not entirely sure how he can spend a whole class period just having students ask questions—but he also knows he will have to help his students rephrase their questions to be more scientific by relating one variable to another. He recognizes that engaging students to ask purposeful, meaningful questions about a phenomenon is an important scientific practice, and he is excited about giving this new approach a try.[1]

To obtain a sharper and richer picture of how PBL can be implemented in U.S. and Finnish classrooms, this chapter and the next detail actual lessons that were taught in each country. One of the major concerns with the new NGSS in the United States and the standards being implemented in Finland is that the material available for teachers to use in their classrooms is relatively limited—though recently a number of organizations, such as the National Science Teachers Association, have been working to bring new science practices and materials to high-school teachers.[2] What policymakers in both the United States and Europe have developed regarding implementation tends to be at a broad level, focusing on themes related to leadership, networks, partnerships, and resources needed for teacher preparation and professional development.[3] These policy guidelines are essential, but at the classroom level teachers often find themselves struggling, especially when available instructional materials are not aligned with country standards. Once teachers have a grasp of the ideas underlying PBL instruction, they need curriculum support to guide them as they get started. This problem also draws attention to the importance of systematic, coherent professional learning experiences that offer strategies for implementing PBL. (More on this later.)

In a traditional physics classroom, students typically spend considerable time solving mathematics problems by using equations that have been given to them. In the case of forces and motion, they might solve problems using the equation $F = ma$ (Newton's second law of motion), but they do not really understand why, or what problem this equation is supposed to be solving.[4] That is, students learn how to plug numbers into the correct parts of the equation, but they rarely understand the reasons behind doing so and what phenomenon this representation describes. When conducting investigations, too, they are guided by a specific text that describes step-by-step procedures for how numbers should be placed in a table or how a graph should be constructed. Students rarely explore how the elements of the equation might interact with one another, or how the equation applies to explaining and predicting phenomena that they actually can experience.

In PBL, students learn the same mathematical relationships or equation. The difference is that by engaging in specific scientific and engineering practices, they can gain an understanding of why this equation is a good representation for explaining and predicting natural phenomena. One common criticism of this type of PBL instruction is that students will not be able to solve the same types of problems that traditionally appear in physics textbooks. We argue that by developing a deeper understanding of how the equations work, students can apply the practices for investigating, constructing, and modeling with data to find the solution to a problem that, in turn, contributes evidence toward making and supporting a claim. These types of activities give students the experience they need to apply these principles to solving other problems.[5]

A Day in Mr. Cook's Classroom

As described earlier, all participating teachers receive extensive professional learning experiences prior to starting the PBL units, as well as virtual support during the weeks they teach their lessons. All PBL teachers are provided with "daily lessonlevel plans" for each unit, which contain lesson learning performances, the driving question for the lesson, descriptions of activities, resources and material lists, and extensive notes to the teacher for facilitating the learning process. Three of these lesson plans (from early, middle, and late in the unit) are included at the end of this chapter. These lesson plans and descriptions of Mr. Cook and his students' experiences are presented using pictures from videos, observations, interviews, and ESM responses of students and teachers.[6] Collectively, these data highlight how the specific design principles of PBL are enacted throughout the unit.

FORCES AND MOTION UNIT: FIRST DAYS

With his lesson plan in hand, and the professional learning session fresh in his mind, Mr. Cook begins his first PBL lesson.

Once all of the students arrive, Mr. Cook's nervousness fades and he begins doing what he loves . . . interacting with the students. He starts the class by asking the students if they have ever been in a car accident, or if they know of someone close to them who has been in a car accident. This begins a discussion among students related to their experiences with car crashes. Mr. Cook warns the students that the video they are about to observe may be difficult for them to watch, and that if they would prefer not to watch it—if they had been involved in a serious car crash, or for any other reason—they should please let him know. He then starts the video, which depicts a crash test with a small car hitting a wall at 120 miles per hour. Students gasp in shock as the car is totaled, smashed beyond recognition. The video replays the crash in slow motion so that students can see the actual impact as the car hits the wall. Mr. Cook then introduces the driving question for the unit: "How can I design a vehicle that is safer for the passenger during a collision?" This driving question will situate the learning for the next two to three weeks.

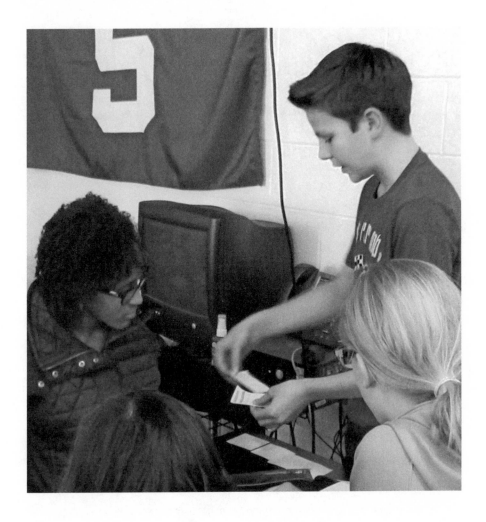

Mr. Cook then replays the video, this time asking students to pay close attention to identify questions they will need to ask in order to determine how to design a safer vehicle. He provides students with Post-it Notes and they begin individually writing their questions. Once each student generates two or three questions, Mr. Cook puts the class into groups and tasks them with deciding how they might organize the questions to determine common themes. There is lively discussion about mass, force, and the velocity of the car. Students are reading each other's questions and commenting as they manipulate the Post-it Notes.

One student asks, "Where should we put questions about the size of the car?" Another student answers, "Everything about the vehicle should go here." Another stu-

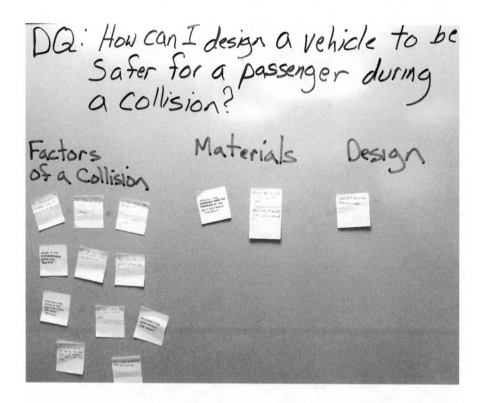

dent, who rarely speaks in class, shyly shares her question, "How can we find out what to fix in each vehicle?" A student responds, asking her, "Would that be about the car or the car design?" The two students agree that the question is about the car design and put the question in the agreed-on category. Students begin to see themes emerging: "It looks like we have three categories . . . questions about collisions, questions about force, and questions about the vehicle." The teacher joins the group and asks questions to promote discourse and critical thinking about the variables that affected the crash. "When you say 'vehicle questions' . . . what do you mean? What about the vehicle? Can you tell me more?"

The class posts each group's questions in a shared space on the room's driving question board.[7] (The picture shown here is an example of the driving question board used throughout the unit to post new questions, student data from investigations, and ideas for car designs.) As the class period ends, Mr. Cook tells the students that they will be investigating the answers to their questions over the next two to three weeks, including building and testing their own safer car design.

During this lesson, we were able to obtain measures of the students' social and emotional experiences as they engaged in various unit tasks. We have found that students are engaged when spending time asking questions, exploring other people's ideas, and solving complex problems—this was evident in the first lesson, when the students came up with their own questions and discussed with others their initial explanations. This also made the questions more personally meaningful and linked the students to the importance and value of the unit's driving question.

FORCES AND MOTION UNIT: THE MIDDLE

Turning now to days four and five in the PBL Forces and Motion unit, the students are working on developing a model to account for the data they have collected: between day one and today, the students have planned and carried out investigations in which they collected data and examined how different variables affect the force of a collision.[8] The lesson plan for day four calls for Mr. Cook to have students develop models to explain the results of their investigation. Although he has used models to teach physics before, he has never asked students to construct their own models to explain phenomena. Mr. Cook reads his lesson plan carefully, because he is unsure of how to support students in this new scientific practice.

Thinking back to the last few days of teaching the PBL unit, Mr. Cook is amazed at the progress he and his students have made. He had always assigned students lab investigations in which the students were given explicit directions to follow. The lab tasks typically occurred after a full class period of lecture, in which he had explained the science principles behind the investigation. This PBL unit was totally different. He is asking his students to design their own investigation without a lecture or direct instruction from him! It was so difficult for him not to just tell the students what to do, and they certainly kept asking him! Every time they asked, he would ask them what they thought . . . just as the facilitators had done during the professional learning session he had attended. Pretty soon, and much to his surprise, the students started to come up with their own ideas and they eagerly went about figuring things out for themselves.

The results were surprising: each group had designed a slightly different investigation. Bringing together their results and sharing them with each other, the students were able to identify the effect of different variables on the force of collisions. This was the first step in understanding and applying Newton's second law of motion.

The next day (day five), when students come into the class, there is an air of excitement. The data and initial models that had been shared the previous day are on chart paper around the room. Mr. Cook has something new displayed from his computer. When class starts, Mr. Cook asks the students to review their data and models from the previous lesson and share with their partner how the data might help them respond to the driving question and design their car to be safer. Students move to their posted data charts and discuss their results. After a few minutes, they share their ideas with the class. Mr. Cook is pleasantly surprised by how students are able to make connections between the data from their investigation and the vehicle collisions. Most groups share that their data showed the need to consider either the mass or velocity of cars in order to design safety features for passengers.

Some students begin to argue over which variable they should focus on in their design. One student suggests that changing velocity is more important because the mass of the car during the collision is not going to change. Another student responds that if

they develop cars with less mass to begin with, the impact of the collision would be less. Mr. Cook is excited to see his students' enthusiasm and how they are questioning and imagining different solutions that would increase the safety of the cars. He suggests to his students that using a computer modeling program would allow them to investigate the impact of varying the intensity of mass and velocity.

Mr. Cook begins to demonstrate for the students the use of an interactive computer modeling program called SageModeler to test their ideas.[9] Before class started, Mr. Cook had been worried that the SageModeler program would be difficult for students to use. As soon as the laptops are distributed, however, Mr. Cook can see that most students are able to use the program without much help. As students develop their models and test them, Mr. Cook notices them eagerly sharing with other students new dimensions of the modeling tool as they discover them. Mr. Cook walks around the class and is very pleased with the progress that the students are making. He notices that the computer modeling program supports the students in thinking critically about the mathematical relationships between the variables without having to do the calculations. He thinks to himself that the modeling tool is particularly helpful for engaging students with lower math skills in mathematical problem solving.

As the class period is ending, Mr. Cook realizes he will not have enough time for all

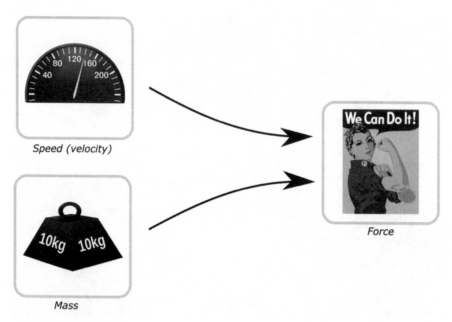

Speed (velocity)

Mass

Force

This is an initial model that captures a basic understanding of Newton's second law, but the students use speed instead of acceleration. This is what a student might get out of a traditional physics class. In PBL, by contrast, students would be encouraged to unpack what this equation really means and how it relates to a real-world application, such as building a safer vehicle.

of the students to share their models. Given how important it is to have students share and critique their models, he knows he needs to make time for this tomorrow. He has also observed that some students were not considering the data from their investigations as they developed their computer models. Recognizing the value of linking the students' investigations with the modeling, Mr. Cook begins planning the next day's instruction—and what principles he will need to emphasize.

FORCES AND MOTION UNIT: FINAL DAYS

The final days demonstrate how throughout the unit the activities have been building on each other, moving students from basic ideas to sophisticated models in a coherent way. Our observations, videos, and ESM show how the students' understanding evolves as they progress through the unit, tackling questions that require the use of big ideas. Our intent is to ensure that every lesson is purposeful and meaningful to the students, captures their imagination, and inspires their sense of wonder.

The two last days of the PBL physics unit—with "How can I design a vehicle to be safer for a passenger during a collision?" as its driving question—have arrived. Students have been preparing their presentations and testing their designs for the past two days. Mr. Cook looks over the lesson plan for the final presentation one last time. He has never had students give presentations in his class before and wants to make sure things go smoothly.

The students come into the classroom and begin to assemble all of the pieces needed for their presentations. They are eager to share their design solutions. Mr. Cook can hear a few students expressing relief that they will not have to stand in front of the whole class to present their designs, since they'll instead be presenting to small groups of their classmates.

Mr. Cook helps the students organize themselves as either presenters or audience members at each table. Each group presents for about five minutes, then the audience moves to the next table. Although there is a little confusion when the students in the audience switch to being presenters, for the most part things go very smoothly. Mr. Cook moves between the tables, listening to each presentation. Most students are engaged and excited to see the teams demonstrate their designs—they are asking each other questions and giving feedback. There are only a few students who do not seem to be participating in the discussions, but they are still listening to the presentations.

The presentations vary in complexity and quality, but everyone is able to share something. Students who did very little work during the previous physics unit are now participating, asking questions and offering their ideas. Although two weeks ago Mr. Cook was not sure about how this unit would work in his classroom, he is pleased with the results he has seen in this final lesson. He observed that his students were really able to use physics ideas to explain phenomena, and he saw that his students were using their creativity and critical thinking skills.

Even though this change has not been easy for him or his students, he feels it was

well worth it as he watches his students' engagement in doing real physics. As the class ends, Mr. Cook begins to wonder if there is a way to use PBL with some of his other physics units.

The presentation of these lesson-level plans is to reinforce the idea that PBL is not about doing a project, like building a model car: it is about asking and responding to questions that are designed to lead to an explanation of phenomena that will in turn build learners' knowledge across the three dimensions. That understanding comes from the students, not just the teacher, asking questions. The students are encouraged to think imaginatively, and all ideas are valued but also open to critique—another important feature of the NGSS and PBL.[10] Each of the six units described in Chapter 1 have these same types of lesson-level plans, which will ultimately be open source and available on our website: scienceengagement. com. (The site will also include materials designed for its Finnish audience.)

Every one of the PBL units has a storyline that the teachers can consult as they move through the lessons. The value of the storyline is to keep the teaching and learning coherent and to connect the activities with their relationship to three-dimensional learning. The storyline begins by targeting the unit's driving question,

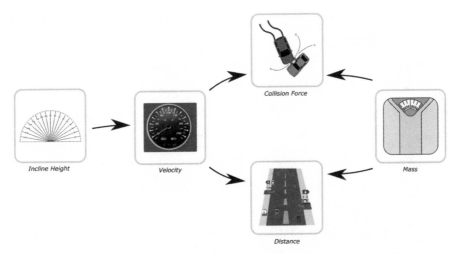

This is an example of a final model that has been updated to reflect the data that the students gathered throughout the unit. While not perfect, it shows that the students recognize which factors contribute to the force of a collision. They have also displayed relationships among velocity, mass, and other variables. In this case, incline height and the distance traveled were not in their initial model but are included here because they encountered these ideas when exploring explanations for the car crash phenomenon.

then it addresses the NGSS performance expectations, which set the learning goals for students. Next there is a section that links lessons through questions, phenomena, scientific and engineering practices, and what is "figured out" and how it is demonstrated within a three-dimensional learning framework. For example, students will generate questions about, and create initial models of, the relationships among force, mass, and acceleration. The unit storyline for the Forces and Motion unit can be found in Appendix B. Due to space limitations, we have not shown a chemistry unit—its basic framework, however, is similar to that of the physics units (see Chapter 1).

Studying Fidelity of Implementation

How likely is it that Mr. Cook's implementation of PBL was carried out by other teachers and in lessons when the teachers were not observed? The lessons presented here were constructed from observational data from three phases of the lessons—at the beginning of the unit, at the middle, and at the end—and from video information. This is the system we used when observing the other teachers in both the U.S. sample and in Finland.

A rubric and scoring mechanism for these observations were constructed to measure the extent to which the teacher addressed the predetermined lesson-level goals; explored the driving question; provided opportunities for students to raise questions about the phenomena being explored; presented accurate disciplinary core ideas and obtained accurate information from students; used scientific practices (asking questions, defining problems, developing models, planning an investigation, analyzing data, interpreting data, solving problems, constructing an explanation, and designing a solution); participated in collaborative activities; encouraged questioning by the students and promoted meaningful whole-class and small-group discussions; and linked student knowledge to prior knowledge and previous lessons. Each of these categories is marked as exhibiting limited evidence, partial evidence, or full evidence. Additional categories were also established to measure the intensity of collaborative work. In Finland, the observation protocol was also conducted in the classrooms, typically by students in the teacher preparation program.

As this work was being conducted in the field test, we also relied on situational data obtained from the teachers and students in the United States and Finland who used the ESM (see Chapter 4 for more on this). These data provided other verification measures of what was happening in the classroom, including a comprehensive list of the scientific and engineering practices that the teachers and students reported using during class. These measures of teacher practices in both Finland and the United States have resulted in several international publications.[11]

The pictures in this chapter were purposively selected to illustrate the involvement of the students in these lessons, especially among females. One might ask, "Well, these are just pictures—how the females were actually feeling is unclear." Yet social and emotional measures retrieved multiple times throughout the semester show that females are more likely to become more actively engaged during PBL units than during their "regular" science lessons.[12] When in PBL, females become more engaged in exploring problems and solving them, and they report feeling more successful than they do in their "regular" science lessons—in both Finland and the United States (see Chapter 4).

All of the teachers in the study participated in the ESM, and were asked in specific ways about which scientific practices they were using in their science lessons. This ESM information, coupled with observations and videos, affords a triangulated view of what actually occurs in PBL compared to other types of science lessons. From this information, a more complete picture emerges of the extent to which Mr. Cook's students—as well as those of the other teachers—are achieving three-dimensional learning in their PBL units. The challenges and successes that teachers are having integrating the PBL framework into their science lessons is described more fully in Chapter 5.

During the first year of our study, a group of Finnish physics teachers visited and observed several lessons in U.S. physics classes. They each commented that the U.S. teachers were all using the same technology program that they use in Finland—one that allows students to immediately give answers to specific questions, then view a score that shows the class's level of agreement on each of the questions. Both the U.S. and Finnish teachers raised concerns about whether this technology for achieving benchmarks of student learning was useful for student understanding and application of concepts. Teachers in both countries expressed a need for more student experiences that would engage them in asking questions and in pursuing investigations that are designed to understand phenomena and solve problems. The classes of Mr. Cook and Elias Falck—whom we will be reading about next—demonstrate how these teachers are attempting to make the transition to experiences that can achieve these goals.

Mr. Cook's Lesson Plan: First Days

UNIT DRIVING QUESTION[a]

How can I design a vehicle to be safer for a passenger during a collision?

Targeted NGSS Performance Expectations

Students who demonstrate understanding can:

HS-PS2-1 Analyze data to support the claim that Newton's second law of motion describes the mathematical relationship among the net force on a macroscopic object, its mass, and its acceleration.

HS-PS2-3 Apply scientific and engineering ideas to design, evaluate, and refine a device that minimizes the force on a macroscopic object during a collision.[b]

Lesson Day	Learning Performance[c]	Project-Based Learning Elements	Materials
1 & 2	Students will generate questions about, and create initial models of, a series of two-body collisions.	**Sub-questions:**[d] What happens during a vehicle collision? **Connection:** This lesson introduces the driving question and engages students in the phenomena of collisions through videos and exploration with toy cars. The driving question will be revisited in lessons throughout the unit. The models built from exploring collisions with the toy cars will also be revised throughout the unit. **Safety Guidelines:** Students should keep toy cars on the surface of their tables or counter. **Assessment or Artifact:**[e] Students will create: • Post-it Notes with questions related to the driving question. • An initial model explaining what's happening during a car crash.	• Post-it Notes • Projector for viewing videos
Lesson Plan	**Teacher notes** Start the unit by asking students if anyone has ever experienced a car crash. Discuss with class. Mention that we will be learning about how to make a car crash safer. Also, keep in mind that some of these topics may be triggers for certain students who have experienced a traumatic event.		

Introduction of the Driving Question

1. The teacher will show students videos of car crashes, then present the driving question and explain the challenge to design a vehicle that is as safe as possible during a collision.

Students will generate questions to determine what additional information they need to respond to the driving question. Students should generate a list of questions on their student sheet.[f] Students will be given three Post-it Notes and will write one question on each (the top three questions that they came up with). Students will attempt to categorize their questions in small groups based on the question topics.

2. The class will come back together and share some of their questions. The teacher will create a list of several categories of questions on the board (based on students' questions and the categories they came up with in small groups). Students will then post their questions in the room on a driving question board,[g] based on the categories. Examples: (1) all questions related to designing a vehicle in one area; (2) questions related to vehicle safety in another; (3) questions related to force in a third.

3. The teacher will emphasize that the class will return to these questions periodically throughout the unit.

Lesson Activities

1. The teacher will look for a question similar to the lesson question or just add "What happens during a vehicle collision?" to the list. This question can be used to introduce the lesson activity.

2. Students will view a series of videos that show collisions between two bodies. The teacher should prompt students to think about the similarities and differences in the factors (patterns) that affect the collisions and why they were (or were not) so violent.

- Van vs. Car (1:30–3:38) 120 mph mega crash (3:09)
- Mercedes S-class vs. Smart Car
- Possible football collisions video. Note: Teacher may not want to show football collisions video, or may want to warn students, as some of the hits are pretty intense. For some students, it will be a good engagement tool; others might disagree.

3. The teacher will ask students to generate a question that they would like to explore related to the factors that affect collisions (this could be a question from earlier, or a new question based on observations from the videos). The teacher tells students to draw a picture of one of the vehicle collisions to explain the response to their question. Students should create a drawing on the back of their student sheet that explains what happened and why. Students should share their picture with a partner to explain their collision.[h]

The teacher should initiate a discussion on models with the class by asking some of the following questions:[i]

- What is a model? What do you think of when you hear the word model?
- What is a model in science? Why are models used? Why are models important?

- Was your drawing a model? Why? Why not?
- How could you change it to make it a better model?
- How are the toy cars different from real collisions? What other limitations does this model have?

Students shouldn't focus on drawing the best-looking toy car, but rather on what's happening during a collision between the two objects to provide an explanation of what happens when a vehicle crashes.

Concluding the Lesson
The class revisits the sub-questions and unit driving question. Students share observations that occurred during the activity and the initial models that they created. The teacher should try to connect the videos from earlier in the lesson to the toy car collisions. Students also share any new questions they have about collisions or that are related to the driving question.

a. Each daily lesson plan restates the driving question that is threaded throughout the unit. It is restated every day to emphasize its importance in anchoring students' work toward explaining a phenomenon.

b. HS-PS2-1 and HS-PS2-3 are the performance expectations identified from the NGSS. They integrate three dimensions of learning instead of aligning only to content. We do not necessarily expect that a teacher will cover an entire performance expectation during a single unit. We have developed these units to assist teachers in working toward increasing students' proficiency in these areas throughout the school year.

c. A learning performance is a three-dimensional learning goal that is smaller in scope than a performance expectation and that students can work on over a day or several days. It helps to build toward responding to the driving question and mastering the performance expectation.

d. Sub-questions help the teacher organize the lesson so that students can work toward responding to the unit driving question by investigating one aspect of it at a time.

e. The assessment or artifact for each day includes a performance goal that helps the teacher to assess students' developing understandings.

f. The student sheet is a document that we have created and provided to the teachers to aid them in eliciting questions from students. Our research shows that they can be especially useful for guiding the students' focus on the driving question.

g. The driving question board is where the class collects and posts all of the questions; in this case, questions about vehicle collisions. The questions on the board are revisited throughout the unit as students answer them. This process gives direction to students' investigations.

h. The activity outlined here is designed to initiate students' thinking about explaining phenomena. We use the word "draw" to introduce learners to the modeling process.

i. Here, the teacher is integrating scientific practices of modeling. These initial models typically show students' growth in their science-related abilities and knowledge as they gather information and revise their models.

Mr. Cook's Lesson Plan: The Middle

Lesson Day	Learning Performance	Project-Based Learning Elements	Materials
4 & 5	Students will create a model to explain the relationship between the force of vehicle collisions and the mass or the speed/acceleration of the objects in the collision, using SageModeler.	**Sub-questions:** How can I develop a system-based model of the collision investigation? **Connection:** Students use SageModeler (SM)[a] to continue developing their models of vehicle collisions. **Safety Guidelines:** None. **Assessment or Artifact:** SM model of the initial investigation of the cart/ramp activity (relationship between speed, mass, and force).	• Laptop cart (1 laptop for every 2 students) • Teacher laptop (to guide students through SM)

Introduction

The teacher introduces the lesson by referring to the driving question; teacher has students share their data and initial models from the previous lesson, describe how their investigation can help them answer the driving question, and get feedback from both the teacher and other students.

The teacher explains that students will be using SageModeler, a systems-based software program available on the web to create a model that can generate data that match what they collected during their investigation. Students will be working in pairs to create a model that explains how the variables from the investigation interact with one another (this may be a good warm-up discussion to get students thinking before they get on the computers).

Lesson Activities

Laptops are distributed. Students create accounts on SM while the teacher moves around the room to troubleshoot. Students should work in pairs for the first day using SM (to limit the technical difficulties that may arise).[b]

The teacher should guide students through the different features of SM using a laptop projected onto the board, going through how to do basic tasks such as create a variable and connect it to another using arrows.

Student pairs should explore the interface for a while (5–10 minutes), until the teacher prompts them to focus on creating a model that can generate data that match what they collected during their investigation. The teacher will move around the room to troubleshoot and check for understanding.

This is the first step in the development of students' models to explain the driving question. Make sure students understand that this model will be revised and changed as they investigate more variables later in the unit.

Sequence:

1. Students open SM link;
2. Students open a textbox and write the driving question;
3. Students start adding variables, connecting them and defining relationships (ask them to add their explanation to each relationship in the reasoning box);
4. The teacher demonstrates how to run a simulation of the model and create a graph;
5. Students run their model to show how the independent variables affect the dependent variables;
6. The class shares and discusses some of the students' models.[c]

The teacher uses the teacher report in the portal to project several student models, then asks students to respectfully critique the various models: What is good about them and what could be improved? What questions do students have for their classmates regarding why they set up the models this way? When showing a particular model, the teacher asks students to predict what the output should look like if one or more independent variables were changed.

The teacher should be on the lookout for common modeling issues, as well. The teacher might want to bring up specific student examples to highlight and discuss—ones that could potentially be improved in the following areas:

- Objects vs. variables;
- Defining proper relationships among those variables;
- Inclusion of appropriate variables—determining what variables have large versus small effects on the system;
- Direct and indirect links between variables—making sure only direct effects are linked;
- Being able to determine the boundaries of the system.[d]

Concluding the Lesson

The class revisits the questions generated at the beginning of the unit and determines which ones have been addressed. The teacher asks what the benefits of having multiple representations of a model or of data from an investigation would be. The teacher also asks: What is the benefit of having a system-dynamic model? Students share the progress they have made with the class (sharing with the teacher via Google Drive).

a. SageModeler is a Creative Commons open-source online tool developed by the Concord Consortium in collaboration with CREATE for STEM at Michigan State University. SageModeler allows students to build their own models, generate data from the model, and use their own and secondary data to test their models and make revisions. Using technology tools to scaffold learning is an essential component of PBL design principles.

b. Working in pairs is another important PBL practice that is designed to have students participate in the type of work scientists use when solving problems. Our data show that these collaborative experiences are highly related to students' engagement and measures of creativity.

c. The goal of these lessons is to have the students engage in discussions, not merely to replicate what others have accomplished. By discussing other students' models, we find students listening more to different points of view and incorporating these ideas into their own models.

d. These PBL experiences are new not only for the students, but also for the teachers. These bullet points help and affirm the performance expectations and scientific practices that the teacher needs to pay attention to.

Mr. Cook's Lesson Plan: Final Days

Lesson Day	Learning Performance	Project-Based Learning Elements	Materials
10 & 11	Students will use models and data from investigations as evidence to communicate and explain the relationship between force, motion, and speed over time and vehicle safety.	**Sub-questions:** How are forces and motion related to vehicle safety? **Connection:** Students create a culminating artifact in the final two days of the unit that seeks to connect one of the lessons to their designed safety apparatus and driving question. **Safety Guidelines:** Students should be monitored on the laptops for appropriate use. **Assessment or Artifact:** Students will create an artifact (PowerPoint, poster, etc.) to accompany their presentation on one of the daily lessons and their safety apparatus.[a]	• Laptops • Markers/ posters (optional)

Introduction

1. Warm-up: What was your favorite lesson during this unit and why? What did you learn in this unit? How does it relate to what we've already learned? The teacher will begin class by asking students to reflect on their learning throughout the unit by sharing their responses to the warm-up questions.

2. The teacher will then go over expectations for final unit artifacts. Students will create a presentation using PowerPoint, Prezi, a poster board, or other medium. Students will present in groups of two. Students will also present and explain their final design solutions during the presentation. The questions below should be given to the students to be answered as part of their presentations.

Lesson 1: What happens during a vehicle collision and why? Or, What makes vehicle collisions so destructive (violent)? Or, What makes one collision more destructive than another? What factors are involved in a vehicle collision? What is a model? Students will share their questions/observations from the videos, as well as their initial models.

Lesson 2: How does force affect a vehicle during a collision? Students will recap the first investigation and share their data and the variables involved.

Lesson 3: How can we use a model to explain the relationship between the different components of a collision? Students will share their graphs, investigation data, and models they created based on the lab. Students will also discuss the role of modeling in science.

Lesson 4: How does one minimize the force on a water balloon during a water balloon toss? Or, How is a water balloon toss related to safety during a vehicle collision? Students will recap the class activity and discuss the concept of *impulse* and how it relates to water balloon catching and vehicle collisions. Students will also share their models from this activity.

Lesson 5/6: How can I develop a web-based model of the collision investigation? Students will discuss the importance of modeling, the use of web-based models in science, and share/ explain the models they created using SageModeler.

Student Activities

3. Students will spend the majority of the lesson working on their presentations and artifacts.

Concluding the Lesson

4. The teacher will field any questions regarding presentations, and check in on students to ensure that everyone will be prepared to present in class the next day.

5. The students should create a claim, use evidence, and develop with their reasoning skills an explanation for why their new design is safer in crashes.[b]

Day 2

Introduction

1. Warm-up: What are some good questions to ask presenters today? What are some things to keep in mind as a respectful audience member? As a skilled presenter?[c]

2. The teacher will begin class by going over the "Carousel" presentation process. It will begin with half of the groups at a position around the room with their presentation artifacts ready and the other half of the class serving as the audience. Students will spend 3–5 minutes at each presenters' station listening to the presentation, asking questions, and providing feedback. Students will switch and move to the next presentation until the audience has seen every group. The groups will then switch and repeat the process.

Student Activities

3. Students will take part in the Carousel protocol, delivering their presentations, sharing their artifacts, and being respectful audience members. Students will submit their artifacts as well as feedback forms they completed as audience members.

a. Students build a vehicle and present it to the class, explaining why their design will be safer in a collision. The expectation is that students' explanations will connect back to the relationships they learned throughout the unit as they responded to the overall driving question. Their explanations should also reflect their understanding of the NGSS performance expectation.

b. Each of the lessons is designed to highlight specific scientific practices that the students should be undertaking when developing their models as well as when they plan investigations; observe; collect, analyze, and interpret data; and obtain evidence to form claims. This framework makes the process of generating explanations more manageable for students.

c. Obtaining, evaluating, and communicating information is a key science and engineering practice. Consequently students need to not only clearly explain and communicate their design solutions to the class, but also critically assess information that other students present.

3. Project-Based Learning in Finnish Physics Classrooms

Mr. Elias Falck has been working as a high-school physics teacher in the Helsinki metropolitan area for five years.[1] He enjoys and values teaching as well as working collaboratively with other science teachers in his school, university, and surrounding area. In early autumn, at the beginning of this school term, we find him looking forward to meeting this semester's physics class and eager to start teaching a PBL unit on forces and motion. A year ago, he participated in professional learning experiences with the U.S. team in Finland and visited the United States as part of the study's international teacher exchange program. Observing in U.S. classrooms, he thought to himself, "This class is so similar to my physics class, it would be intriguing to develop our own PBL lesson plans." Upon returning to Finland he conferred with a few colleagues, and before long a team of four other teachers from three different schools and one university researcher was assembled to create an innovative PBL physics unit for Finnish secondary schools. Reflecting on last spring's sessions with colleagues, he remembered how they strategically built their unit to both exemplify the Finnish Core Curriculum reforms and overlap with the design principles of PBL.[2]

With the design team's conversation in mind, the Finnish teachers decided to prioritize encouraging students to collaborate and to engage in modeling and creating artifacts, which they agreed otherwise occurred only occasionally in their classrooms. As for the learning performances for the Forces and Motion unit, the team decided to begin by introducing experiments to explain variations in velocity, first for different falling objects, then later for objects colliding on an air track. The planning team agreed that the Finnish students would be prompted to ask scientific questions after viewing a teacher-led demonstration in which two coffee filters with different masses were dropped from the teacher's hands. But most importantly, in keeping with the design principles of PBL, the teaching modules would include opportunities for planning investigations, making evidence-based claims, and communicating findings.

Once again, Elias looked over the lesson plans to make sure that he had everything he needed for the day's class. Slightly anxious, he still was not entirely sure how to initiate the process where students are encouraged to ask questions—especially because he was accustomed to starting lessons with short presentations about principal topics and with questions he had come up with, not those generated by his students. Previously, he had tried to initiate a student-led questioning process, which did not go well—but then again, this earlier effort was not part of a PBL unit. Thinking to himself, he realized there were several reasons that prior initiative did not go as planned, and that this new effort was very different in its unified questioning process, coherence among ideas and activities, and assessments, which included developing computer models. Even though he knew these new lessons were going to move him out of his comfort zone and be personally challenging, he was ready and committed to the PBL framework, which involves engaging students to ask purposeful meaningful questions about a phenomenon.

In this chapter, we describe two ninety-minute lessons in Elias's physics classroom as he tackles teaching a Finnish-designed PBL unit on Forces and Motion. In contrast to many U.S. high-school students, Finnish students are already familiar with basic Newtonian mechanics when they enter high school; they learn about concepts such as motion with constant velocity and acceleration in middle school. Other topics covered in middle school include the cause and effect between a force acting on a body and the change in velocity or acceleration. By high school, students have been exposed to the effects of forces like gravity, friction, normal force, and air resistance. Unlike secondary students in the United States, every Finnish high-school student is required to take a physics course and become familiar with two basic models of motion—constant velocity and constant acceleration—using graphic representations and equations.[3] The new Finnish Core Curriculum for both lower and upper secondary school identifies these goals as quite similar to the U.S. performance expectations specified in the NGSS.[4] Indeed, the enactment of the NGSS and its emphasis on doing science is consistent with the new Finnish curriculum, which highlights the importance of inquiry and use of scientific practices—skills that Finnish educators and stakeholders perceive as critical for the twenty-first century.[5] According to the Programme for International Student Assessment (PISA), which asks students about the scientific practices they use in their science classes, Finnish students report spending less time asking questions and planning investigations than do students in other OECD countries, and they rarely use evidence to draw conclusions.[6] Instead, Finnish students engage in solving textbook problems using mathematical equations—and like students in the United States, they typically do so without understanding what the equation represents.[7] These results have led to the launch of a national project to reinvent primary and lower secondary education practices in Finland.[8]

Finnish teachers were willing to experiment with designing and teaching PBL units in large part because they view themselves as lifelong professional learners and they are interested in developing lessons that are more student- and inquiry-led instead of teacher-led, an idea articulated in new Finnish national guidelines.[9] A fuller description of Finnish teachers' orientation to their practice is discussed in Chapters 5 and 6. The focus here is on the complementary approaches to teaching science in both countries. PBL describes strategies for having students recognize a phenomenon and then figure out why it occurs, where the problem is situated, what variables describe the properties of the phenomenon, and how the phenomenon could be represented (including but not limited to its standard equation) and used to estimate relationships among the variables. These concepts accentuate inquiry and research, which are of primary importance in the new Finnish upper-school curriculum. Understanding what problems need to be solved and what meaning they have in the students' lives are central issues that are fundamental not only to science education, but also to the Finnish education system more generally.[10] Reform in science education is seen as a conduit for developing solutions and pedagogical innovations, a critical component for future Finnish researchers as well as teachers.[11]

It is important to bear in mind that Finland is not adopting, but adapting, PBL principles in the context of their science learning environment. There are cultural differences between Finland and the United States that are reflected in their curriculum structures, and what is reported here is a Finnish interpretation of PBL that is consistent with the aims of Finnish science education. In examining the lessons that follow, it is essential to note those design elements of PBL that are implemented differently in Finland than in the United States.

As in the United States, the Finnish teachers participated in professional learning activities. During several country exchanges, teachers also met with PBL experts, including several lead physics and chemistry teachers from the United States. Additionally, there were several video exchanges between the teachers from the United States and Finland. But the most extensive professional learning occurred independently in Finland, with Finnish science educators and researchers working collaboratively with Finnish teachers on the design, daily lesson-level plans, and assessment practices. With respect to assessment, students in Finland do not participate in the same testing regime as that found in the United States. Consequently, Finnish teachers found PBL's formative assessments and the construction of artifacts affirming and consistent with their evaluation practices.

In Elias's PBL Forces and Motion unit, Finnish students are directed toward finding reasons for changes in motion (cause and effect) when two forces act on an object, and then explaining why the changes occurred. As in the U.S. lesson-level outlines, learning goals for tasks are identified as learning performances, and all tasks are connected with a driving question that is threaded throughout the unit. For the Finnish unit, the driving question is "Why do some objects take different amounts of time to fall from the same height?" We have annotated specific terms that are unique to Finland. The lessons, pictures, and text are taken from videos and data collected from Elias's classroom.

Elias's Forces and Motion Unit: Getting Started

Elias starts his lesson by introducing the topic: "This week we will focus on different kinds of motion and the reasons behind them. We will design experiments, discuss them,

and create a product based on what you learn during this project (a project-based learn-
ing activity). To understand the driving question and pose relevant research questions,
let's start with a demonstration. I have two coffee filters. The first one is a single filter
paper and the second consists of two papers stuck together. Let's do a test. What hap-
pens when I let the objects fall? And look carefully at what happens!"

Elias lets the objects fall. The heavier object is the first to hit the ground. He asks,
"What did we observe? Recall what you saw." Janna answers, "The object with two fil-
ters was falling more quickly." Elias continues the demonstration. Now he has four filter
papers stuck together in his right hand and two filter papers in his left. The mass of the
object in his right hand is double the object in his left hand. Again, he asks the students:
"What happened?" The students are unable to distinguish between the two trials.
Next, he holds an object with eight filter papers in his right hand, and in his left hand
an object with four filter papers. Finally, he does a trial with thirty-two and sixteen filter
papers. During the last two tests, the students recognize that the difference in the time
it takes the coffee filters to fall is not as long as in the first test, although one falling object
is still two times heavier than the other. The students are astonished. They wonder why
the falling times of the objects became closer, even though the mass of the second ob-
ject was still double the first. This was not what the students expected would happen.

The demonstration pushes students to wonder and become curious: an excellent starting point for the study of forces and motion. The driving question becomes more understandable and supports students' thinking, helping them pose relevant research questions for the next phase of the lesson. Students are asked to write down questions related to the motion of a falling object and questions related to the reasons for the nature of the motion. Students start working with the questions that they will analyze and discuss together.

Elias summarizes the experiences: "This was the orientation to this week's unit. Later, we will explain what we observed. You might be wondering now, so let's start studying the phenomenon. However, we have to break it down and study it part by part based on your questions. This is not a simple phenomenon. There are many issues. You should have this phenomenon in your mind while we are answering the driving question." Elias asks the students to generate relevant questions related to the falling of the objects, which will support them in answering the driving question. The students generate a list of questions in a small group. After the group activity, the class comes together and shares some of their questions. During the discussion, the questions are divided into two groups: questions related to the motion and questions related to the change in motion.

Elias continues the lesson and shows a video clip of a rocket leaving Earth. He explains that this phenomenon deals with similar issues as the driving question and asks the students to compare the rocket phenomenon to a ball that is thrown toward the sky. The students are asked to come up with their own questions related to motion, such as "Is the velocity constant?"; "How does the motion change or not change?"; and "How does the velocity change, if it changes?"

Referring to the students' questions, Elias says: "Now we will engage in designing experiments according to your questions. Here, we have equipment. You are free to use the microcomputer based laboratory (MBL) tools—for example, the ultrasonic sensor could be useful." The students start planning their investigation according to their questions. The investigations are focused on modeling the motion of different falling objects. While the students are planning and working, Elias walks around the classroom. He asks

questions to help students focus on relevant topics: "How do you feel when an elevator goes up? Or when it just starts to go up? How do you feel when it starts to go down? Please, think about what you feel inside your body—what happens—why?"

This type of questioning helps students orient themselves to the phenomenon and recognize the link between the net force and change in motion. After taking measurements, students spend a long time negotiating, modeling, and communicating. The models they create represent two different types of motion (one with constant velocity and one with constant acceleration) and include preliminary explanations for the changes in velocity. Students also discuss whether the model actually captures the phenomenon and discuss what might be missing, based on the following kinds of questions: Are there things in your model that are not part of the cause and effect relationship? Or does time or some other variable change the explanation (for example, a falling object may only be moving with constant acceleration for a short period of time in the beginning)?

When students are modeling, they have to verify their measurements with more measurements, a process that makes the model more accurate. Recognizing the need for more data to draw better evidence-based conclusions, students drop the object from greater heights and record the time to collision. Observing patterns in their data will help make their claims regarding the phenomenon clearer.

Elias remarks to the class, "Our driving question—'Why do some objects take the same amount of time to fall from the same height while others do not?'—has been partly solved." The students have recognized that heavy objects accelerate the entire time while falling and light objects accelerate for only a short distance. But the reasons for the differences in motion are not yet fully clear.

Elias's Forces and Motion Unit: Next Steps

During the next set of lessons, students will apply their models to more complex situations. In prior lessons in the PBL unit, students had developed models to explain the change in motion among objects of varying mass. In Lessons 4 and 5, they will try to extend their models to describe interactions between multiple objects when they exert forces on each other.

Elias tells the students that, so far, they have analyzed several different kinds of motion: motion with constant velocity, constant acceleration, and motion where the velocity changes but not in a linear way. "Today, we will look in more detail for reasons behind the change in motion. Why does an object change its motion? Why does it start moving or stop? Here, we have a heavy and a light cart." Elias shows all the equipment. He says that he will use MBL tools, which the students are already familiar with, in the demonstration. "We will measure, as earlier, time and velocity. We need volunteers to complete the experiment." The volunteering students push the heavy and light carts toward each other at the same time so that they crash and bounce backward—a process they repeat several times. The MBL tool measures the velocity of the carts and plots the points on a graph, showing that the heavy cart moved back more slowly after the crash than the lighter cart.

Elias tells the students, "Now, get into small groups and start modeling the phenomenon. You should take your whiteboards and start to analyze the graphs." The

students already know how to start, so they begin working in small groups, discussing the phenomenon and making drawings on their whiteboards (constructing a cognitive artifact). Elias walks around the class and supports the students' collaboration. He does not give out the answers but asks small guiding questions like, "What is your argument?" and "What are your data, or what is your evidence/claim?"

After the modeling activity, Elias starts a whole-class discussion: "Why do the graphs of the heavy and light carts look different? Please, explain your answer." Students justify why they think the velocity graph of the heavy cart looks different. Elias continues the experiment and says that they will now use the models developed during the previous experiments. He introduces the air track and air-track gliders and adds repelling magnets to both gliders. Two volunteer students push the gliders and let them collide. The velocity of the gliders is measured before and after the collision. "Let's have a look at the graphs in small groups. Please make a model which describes the reason for the change in velocity." Elias guides the students to recognize situations where the motion changes and why it changes. He asks questions in each situation, like "What happens to the motion?" and "What makes the change happen?"

The aim of these lessons is for students to recognize that there is always an interaction between two bodies when a force is applied. The concept of force is introduced through the collision phenomenon; the task for students is to analyze what happens when two objects are in interaction with each other. In the collision experiment, students are expected to articulate why these two forces result in this interaction: both objects act on the other, and the forces are equal in size but opposite in direction. The teacher then guides students to explain why an object with a larger mass has a smaller acceleration in a collision.

"Let's test how these models explain the phenomena around us." Elias starts with a set of simple demonstrations and asks several questions. Finally, in small collaborative groups, the students start to solve the problems that Elias introduced through the demonstration. At the end of the lesson, he leads a discussion about possible solutions. "Please, present a hypothesis about what is happening in a situation where we have two gliders on an air track. Between the gliders there is a spring and I hold the gliders close together. What happens when I let the gliders go free?" The students present hypotheses and construct arguments about what they think will happen. Elias comments, "Okay. Let's have a look at what happens. You were right. Both gliders feel the same interaction. After the interaction, the velocity of the heavier cart was half that of the lighter cart. Let's continue the analysis of different situations, which I will demonstrate."

The lesson continues by analyzing different motions and reasons for the motions. Elias shows a video clip from the space center, in which an astronaut does various demonstrations in space. The students discuss them in small groups.

It is interesting to note that both Mr. Cook and Elias were anxious about starting PBL—in fact, we found this to be the case with all of the teachers. Once they got into the lessons, however, they were fully engaged and became more comfortable having the students take a greater role in learning science. There is another point that also needs to be raised about Mr. Cook and Elias: they are both male teachers, which in the context of this study needs some explanation. There is first the problem of representation, which is more of a problem in the United States than in Finland; in the United States, males continue to outnumber female teachers, especially in physics.[12] Fortunately, in our sample, both countries include many male and female teachers in chemistry and physics. In fact, in Finland, there are more female than male teachers leading physics and chemistry classes.

Increasing and sustaining females' interest in science is of major concern in both the United States and Finland, and we too have an interest in promoting a more diverse pool of potential scientists in multiple fields. Females in our study, while having lower levels of engagement than males at the start of our PBL units, had greater gains in engagement than the males by the end of the units. Males' average level of engagement, however, remained higher than that of the females. The pattern for interest is different, with males showing a greater growth in interest. At the beginning of the study, the female teachers reported a slightly higher level of connection between the importance of science and their sense of self, but males had a higher increase over the study period, so that their level of connection eclipsed that of the females by the end of the units. This same pattern occurs for perceiving the value of science for the future: males' interest was initially lower than that of females, but it increased more than the females', becoming higher by the end of all of the units (though the difference between males and females on this issue is not as great as that for sense of self).[13] We continue to consider these gender differences in all of our analyses and they are a primary consideration in the design of the PBL units and their assessments.

Studying Finnish Fidelity of Implementation

As in the United States, when testing the impact of the PBL unit in Finland on science learning and social and emotional experiences, it is critical to learn if the teachers actually taught the units in ways that were consistent with PBL principles. To gauge the fidelity of implementation in Finland, data were obtained from classroom observations, selected classroom videos, and administration of the ESM surveys. Analyses of our data from both countries show students situationally engaged in PBL, especially when developing models. Chapters 2 and 3 detail only one of the physics units, but we have found positive effects overall on students' engagement with science for the suite of PBL units.[14] Although teachers and students were often hesitant in the beginning, they soon found PBL engaging. Moreover, preliminary assessments point to the PBL units having enhanced science learning.[15]

Compared to students in Finland, a smaller proportion of U.S. students takes physics, the classes are typically taught as the last science course in high school, and the share of highly advanced students in U.S. classes is just more than a third.[16]

Our observations and analyses of the ESM show that in all of the classes, participatory collaboration in PBL among all genders, races, and ethnicities is surprisingly equal and one of the benefits of PBL. Participatory collaborative teamwork—where everyone is included with shared responsibilities—and "thinking outside the box" are highly valued. But while the teenagers in the Finnish and U.S. classes look similar, there are still major cultural differences between the classrooms. These are perhaps most pronounced in the teachers' training expectations, which we detail in Chapter 5.

Teachers in Finland are well respected—considered experts in their profession—and problems with classroom management and organization are less noticeable than in the United States. In some ways, the Finnish teachers in this study were unusual because they were willing to try not only a different instructional approach but also a total reorientation and reconceptualization of their teaching. For teachers in Finland, who spend considerable time lecturing and asking questions of their students (much like many science teachers in the United States), PBL was indeed challenging. The Finnish teachers made a great effort to not only take on a different instructional approach but also to work bilingually with their U.S. counterparts. Everything was initially prepared in English and then translated into Finnish. Further, once the Finnish videos had captured the narratives, all instruments and measures were then translated back into English.

What the lessons from Finland show is that the design principles of PBL resonate with the mindset of Finnish teachers interested in changing their practice. The willing interest of our Finnish colleagues, and the care with which they implement many of the PBL concepts, suggest that the ability to change instructional practice to one that is more problem-oriented and based around participatory collaboration is not just rhetorically promising, but also possible to implement in an international context. What we are seeing here are the seeds for a possible reform: building on these initial results should help this transformative approach reach more classrooms throughout the United States and Finland.

Elias's Lesson Plan: First Days

UNIT DRIVING QUESTION[a]

Why do some objects take different amounts of time to fall from the same height?

Targeted Performance Expectations According to the Finnish Core Curriculum Framework[b]

Students who demonstrate understanding can:

- Analyze data on the motion of a falling object and recognize when the object moves with constant velocity or changing velocity.
- Introduce models for motion with constant velocity and constant acceleration.
- Analyze data to support the claim that Newton's second law of motion describes the relationship between the net force on a macroscopic object, its mass, and its acceleration.
- Apply scientific and engineering ideas to design, evaluate, and refine an experimental design that could be used for modeling motion with constant velocity and constant acceleration and that helps students recognize the relationship between the net force on a macroscopic object, its mass, and its acceleration.

Lesson Day	Learning Performance[c]	Project-Based Learning Elements	Materials
1 & 2	Students will generate questions about falling objects with different masses.	**Sub-questions:** What happens when two objects are allowed to fall from the same height? The first object is one filter paper and the second object consists of two filter papers that are stuck together. **Connection:**[d] This lesson introduces the driving question and engages students in the phenomenon of falling objects through demonstration, videos, and exploration with falling objects. The driving question will be revisited in lessons throughout the unit. Initial models will be built based on the motion of different falling objects. **Assessment or Artifact:** Students will create: • Graphical models that describe motion with constant velocity and motion with constant acceleration.	• Coffee filters • MBL tools • Projector for viewing videos • Learning management system where each group of students has space for writing down questions, drawing graphical representations, and creating models

		• Initial model/models explaining the net force acting on a falling body and the motion of the body.	

Lesson Plan	**Teacher Notes**
	Mention that this week we will focus on different kinds of motion and reasons behind the motion. Mention that we will start with a classification exercise and then continue with a demonstration: falling coffee filters.

Introduction of the Driving Question

1. Start classifying different motions inside and outside the school building in a small group. Ask students to give examples of their own and others' small and big body motions. The motions could be classified in many ways: linear-curvilinear; motion with constant velocity–motion with changing velocity; linear-vibrations; linear-circular. Ask students to give examples of different situations where the velocity of an object changes.[e] Encourage them to give examples of objects with different masses. What changes the velocity? How does the mass of the object influence the change in velocity? Discuss first in a small group and then with the whole class.

2. Continue with a demonstration in order to support the understanding of the driving question and the posing of relevant student questions. Introduce the falling objects (coffee filter papers). The first object is one filter paper and the second object consists of two filter papers, which are stuck together. The mass of the second object is double that of the first object. Ask what happens when the objects are allowed to fall. Continue the demonstration with the following number of coffee filters: 2–4; 4–8; 8–16; 16–32. Ask students to make a summary of the outcomes of the demonstration. The next phase, posing questions, is based on this summary.

3. In small groups, ask students to generate questions about what additional information they need to answer the driving question. Students generate a list of questions using the learning management system.[f] The students are supposed to ask questions related to the motion of a falling body and reasons behind the change in motion. Guide students to categorize their questions. Ask students to copy the questions to the common space in the learning management system.

4. The class will come back together and share some of their questions. A list of several categories of questions will be created in the learning management system (based on students' questions and the categories they came up with in small groups).

5. The teacher will emphasize that the class will return to these questions periodically throughout the unit. First they will analyze motion and change in motion. During the third lesson they will explain why the motion does or does not change.

Lesson Activities

1. The teacher shows a video clip of a rocket leaving Earth.[g] The teacher explains that this phenomenon actually also belongs to the topic the class will analyze. The teacher asks students to compare the phenomenon to a ball that is thrown toward the sky.

2. Students are guided to design experiments according to the questions they posed. The students select questions related to different types of motion. Questions related to the reasons for the change in motion or velocity will be analyzed later. The teacher introduces the equipment available, like microcomputer based laboratory (MBL) tools.[h]

The models created during the activity will explain two different types of motion (motion with constant speed and motion with constant acceleration). The students will also discuss the validity of their models.

There are several possible representations that could be used for the models: graphs of time versus speed, velocity, or acceleration; algebraic representations or equations like $s = vt$ or $v = at;$ and written descriptions.

The teacher supports students while they work on the modeling activities by having them provide evidence behind the model and by discussing the models, asking questions of the students:[i]

- What is your model?
- What is the representation you use for the model? Could you use another type of representation?
- What is your research question?
- What is your experimental design in order to find answers to the questions?
- What are your data? What is your evidence?
- What do you claim? Is your evidence supportive of your claim?
- How is the model based on your data?

The teacher guides the students to discuss models in general:

- What is a model in science? Why are models used? Why are models important?
- What are representations? Why are different representations needed and used?
- Was your table/graph/drawing a model? Why? Why not?
- How could you change it to make it a better model?
- What makes a model valid?

Concluding the Lesson

The class revisits the unit driving question and what was recognized about students' questions. Students share their models related to the motion of a falling body and present possible reasons behind the following question: Why did the motion change (velocity)?

The students present common models for motion.[j] In one model, a falling object falls with constant acceleration (velocity increases evenly). The net force is based on the weight and/or size. The second model describes falling when the velocity is constant. The weight is equal to air resistance. The first motion could be recognized at the beginning of a heavy object falling. The second motion could be recognized at the end of a light object falling. The model for the first motion is motion with constant acceleration (a = constant) and the model for the second motion is motion with constant velocity (v = constant). The graphical representation for the first motion is a linear line in tv-coordinates and a representation for the second motion is a linear line in ts-coordinates. Moreover, the motion of the falling coffee filter after the beginning and before the constant velocity is motion with changing velocity.

Forces are needed for the model, which describes reasons behind the changes in motion. The conception of force is already, at some level, familiar to the students. In a case where the net force acting on the object is zero, air resistance equals the object's weight and the motion is not changing (i.e., velocity is constant). In a case where the weight is larger than air resistance, the velocity of the falling object is increasing. These models could be presented using an array of representations.

a. One of the major differences in Elias's class now, compared to how he traditionally taught, is that he begins with a driving question and uses it continuously throughout the unit.

b. Rather than being a list of content standards, the new Finnish Core Curriculum framework describes a set of learning goals that include the use of scientific practices integrated with big ideas.

c. A learning performance is a smaller-scale version of the module aims that students can work on in this lesson or several others. It helps to build toward answering the driving question and mastering the module aims.

d. In the U.S. example, we used a car collision, but in Finland, students do not drive—so the phenomenon of falling objects was selected instead as a way to think about net force. The key point here is that students are being asked to explain a phenomenon—in this instance, the difference in acceleration of falling objects—that they do not understand but encounter in their lives.

e. In the first experiment there is a clear difference in the time it takes for the two objects to reach the ground: the heavier object falls faster. In the last experiment, however, the objects take the same amount of time to reach the ground even though the mass of one is double the other. This helps to raise more questions about the phenomenon that students can investigate. It is a good starting point for having students move to more sophisticated questions and their own experiments.

f. The learning management system is commonly used in Finnish science lessons and there is a space in all student groups for posing questions, collecting graphical representations, and sketching models. In the space there are also hints for posing questions, like "What do you already know about the topic?" "What do you want to learn through investigations?," and "What did you learn during the investigations?" These hints are useful for guiding the students' focus on the driving question and for creating models.

g. Students will revisit their questions after seeing another phenomenon that deals with forces and motion. The video is used to spark questions and explanations that they wouldn't have thought about yet (after having seen only the coffee-filter demonstration).

h. MBL tools are electronic sensors, such as an ultrasonic motion detector or accelerometer, that students can use to collect data that would be difficult to collect by hand. This allows them to investigate additional variables—one of the key components of PBL.

i. This is a good example of some of the questions and tasks that Finnish teachers ask of their students. It is common for Finnish teachers to walk around the classroom when students are engaged in experimentation. Asking students questions about their work is a type of assessment that is emphasized in training programs for Finnish science teachers.

j. This explanation is similar to what a Finnish teacher would receive in a lesson. As shown in the example, students can graph the velocity versus time or acceleration versus time of the falling coffee filters to better understand how the mass of the object and air resistance affect its motion. The coffee-filter demonstration is a good "anchoring phenomenon" because it raises questions and provides students with an opportunity to use data to investigate it.

Elias's Lesson Plan: The Middle

Lesson Day	Learning Performance	Project-Based Learning Elements	Materials
4 & 5	Students will create a model to explain how the forces come from the interaction between two bodies and how these forces act on objects and change the motion. The model will explain the influence of the mass of the object on the change in motion.	**Sub-questions:** How does force act on an object to influence changes in the velocity (acceleration) of an object? How does the mass of an object influence changes in velocity (acceleration) when the object is pushed? Where are the forces acting on the object coming from? **Connection:** Students develop their models collaboratively. **Safety Guidelines:** None. **Assessment or Artifact:** Students create a model that describes the motion of an object when force is applied to it.	• Heavy and light cart • Heavy and light air-track glider • MBL tools (interface and ultrasonic sensor) • Air track and two gliders (the second has double the mass of the first)

Introduction

The lesson begins by referring to the driving question. Students will analyze velocity data measured in two situations: (1) change in velocity while pushing a heavy and light object; and (2) change in velocity while a heavy and light object collide. Students will be working in pairs to create a model that explains how the variables from the investigation interact with one another.

Lesson Activities

1. Students push heavy and light objects and measure the change in velocity with an MBL tool. Students construct a model that explains the influence of the mass of the object on the change in velocity, then share and discuss their models together.

2. The concept of "force" is approached through analyzing collisions of two air-track gliders. The mass of the second glider is double that of the first one. This experiment starts with a discussion about collisions based on a couple of video clips. The demonstration and discussions that support the students' learning underscore the following:

- In a collision, two objects interact and two forces result from this interaction. Each object acts on the other: forces are equal in size but opposite in direction. These two forces are called action and reaction forces.
- The object with a larger mass will have a smaller acceleration.
- In a situation where one object is big, like a wall or the Earth, the change in the motion could be recognized only in the motion of the smaller body. The forces are the same, but the acceleration of the big object is very small.

Concluding the Lesson[a]

The class revisits the questions generated at the beginning of the unit and determines which ones have been addressed up to that point. The class will revisit the lesson's learning performances. The teacher asks, "What are the benefits of having multiple representations of a model or data from multiple investigations?" Also, "What is the benefit of having a web-based model?" Students share their progress with each other and what they have made in class online.

a. The questions related to the models describing motion have been analyzed, but the questions related to the reasons for the change in motion have not yet been answered. They occur later in the unit.

PART II

Measuring Science Engagement and Learning

4. How Learning Science Affects Emotions and Achievement

Why is the cost of a liter of water in some countries higher than that of a can of soda? Are there other planets outside our solar system that could support human life as we know it? There are millions of scientific questions that can pique adolescents' interest. When presented with questions that one finds personally meaningful and intellectually challenging, everyone—teachers, students, and parents—can become more interested in science topics and want to learn more about them.[1] But meaningful questions that could advance science knowledge are not being asked in many classrooms. A recent survey among U.S. scientists finds that 84 percent of them believe that K–12 STEM education in the United States overwhelmingly fails to link science learning with students' interests and make the case that it is important to their future lives.[2] Encouraging students to learn science requires environments that ignite students' interest in asking meaningful questions, tackling challenging problems, and acquiring new skills. Results from the first phase of our study show that students in both Finland and the United States are more likely to grow in their imaginative and problem-solving abilities when engaged in positive science experiences.

PBL principles are particularly well suited for advancing science learning, because they are organized around purposeful and meaningful questions. Although "questioning" is widely recognized as a motivator for enhancing interest in science and as a catalyst for the types of experiences that can actualize academic, social, and emotional learning, few studies have examined these ideas in high-school science environments. Yet reports by the National Research Council, as well as the Next Generation Science Standards (NGSS) and the Finnish Core Curriculum, are based on the assumption that by identifying performance expectations (or "competencies" in Finland), and specifying how they should be taught, the level of interest in science by students and teachers, as well as how they work on scientific problems, can be substantively changed.[3] The driver for our entire project is to learn if, with a PBL-principled system designed to advance science learning, we can help to transform high-school science. Toward this end, we have been assessing how PBL science environments affect students' social, emotional, and academic learning.

Situational Nature of Emotions

National reports and scholars emphasize that one problem with science learning is that students are not engaged when in science class. Why? Much of life in school is fluid, and students' social and emotional experiences vary throughout their daily classroom activities. We do not expect students to be fully engaged in their schoolwork all the time any more than we expect adults to be fully engaged in the same activity at home or at work for hours on end. In fact, some might say that adolescents' earlier stage of neurological development makes it harder to motivate and engage them in activities than adults.[4] There is a high bar for motivating adolescents to become meaningfully engaged. Students find some classroom activities worth paying attention to and they enjoy doing them. Other tasks, such as copying how to solve a problem from a whiteboard, are less captivating, and so are likely to increase disinterest and boredom.

Although researchers agree that engagement is a changeable experience that varies over time, many studies pay only limited attention to what happens to students in their everyday learning contexts.[5] To obtain measures of students' engagement experiences and other subjective feelings, researchers traditionally employ surveys that assess these conditions retrospectively. This approach, however, often fails to capture both the variability in how students feel from one moment to the next and the context(s) in which that variability is situated. Recognizing the difficulty of trying to define engagement without specifying when it occurs misses being able to identify moments when students are actually involved and feeling successful at what they are doing. We argue that engagement in science is situational: not all experiences have the same effect on social, emotional, and academic learning.

Further, our approach identifies a particular set of constructs that are critical for enhancing student engagement when students are in these specific situations. First is interest: that is, an experiential moment when students are faced with a problem or phenomenon that is relevant to their life—for example, how can we build a safer car? Second, to solve that problem, students have to believe that they have the skills needed to begin creating a solution. And third, whatever the students' skill level, finding a reasonable solution should still be a challenge. It is not easy to balance skill and challenge, but doing so remains key to student engagement.

Engagement, however, is dependent not only on the subjective feelings of interest, skill, and challenge, but also by how these feelings play out within a particular situation. Further, situations can be designed specifically to enhance these subjective feelings.[6] We call enhanced situations like this optimal learning moments (OLMs). When students are fully engaged in these OLMs, they are more likely to feel positive about their work, acquire new knowledge, use their imagination, and stretch their problem-solving abilities.

The methodology we use in our work allows us to measure with precision how students feel in a particular activity and what they are doing in real time, which enables us to identify moments of situational engagement both in and out of the classroom. To capture these instances, we use the ESM, a form of time diary survey designed and used by the well-known social psychologist Mihaly Csikszentmihalyi to provide evidence for his theory of "flow": that human experience when a person is so involved in a task that it feels as if time has flown by (or, as others have described it, when one is in the "zone" of optimal performance).[7] In our experience, these are times when students are so deeply engrossed in an activity that they continue working on it, regardless of time constraints—for instance, continuing to work instead of rushing out the door once the bell rings. These feelings can now be obtained and recorded with smartphone technology that randomly prompts students to answer a set of academic, as well as social and emotional, questions throughout the day.

Because we want to both capture and contextualize these moments, the ESM offers advantages over other, one-time surveys related to social and emotional learning. Rather than relying on a scale that is administered only once, we are able to capture even subtle variations in adolescents' thoughts and feelings as they interact within specific contexts, both in and out of school: for instance, when doing an experiment in science class rather than watching a video at home with their family.

The ESM has been used in a variety of studies on adolescents, including those that describe behavior or seek to understand the experiences that students have in different situations.[8] For example, Lee Shumow, Jennifer Schmidt, and Hayal Kackar have studied adolescents' cognitive, affective, and motivational states while doing homework, and have found that contextual factors—such as who the students were with and whether doing homework was their primary activity—were associated with self-esteem and grades.[9] ESM data about moment-to-moment, situational experiences can also be helpful for teachers trying to assess how students feel during specific activities. With near-immediate feedback, teachers can learn which task situations students find too challenging or unimportant to their futures and modify the classroom experience accordingly.

Our smartphones are programmed to deliver signals at random times throughout the period of study—usually over the course of several days—to capture the variety of experiences the students encounter. When students receive the signal, they answer a short series of questions on their phones, a process that usually takes only a few minutes.[10] This information is then uploaded to a secure server; within a matter of days, it is ready for analysis. Previously designed graphs and tables that show information on classes can be retrieved to provide almost instant feedback on instruction and classroom experiences. We have used several of these graphs in our professional development experiences to demonstrate how the data can help identify particular activities that increase engagement, imagination, and problem-solving.

Measuring Social and Emotional Learning in the Moment

What, specifically, do we measure with the ESM? As explained earlier, at the center of our OLM model are what we consider the key constructs of engagement—that is, times when one's interest, skill, and challenge are all above their average levels. (From a social psychological perspective, "interest" is the psychological predisposition for a specific activity, topic, or object; "skill" is the mastery of a set of specific tasks; and "challenge" is the desire to take on a difficult, somewhat unpredictable course of action.) When fully engaged, we expect to see students concentrating, feeling in control, and reporting that they feel as if time is flying by. To test the strength and relationships of other emotions when engaged, we also asked about feelings of being excited, proud, cooperative, competitive, and lonely.

When fully engaged in a learning task, students typically experience some or all of three other related subjective experiences. The first of these experiences—in which positive emotions that we call "learning enhancers" are activated—happens when students are enjoying what they are doing, succeeding at what they are doing, and feeling happy, confident, and active. The second—"detractors"—are experiences that students are unlikely to have when engaged, such as feelings of boredom or confusion. And the third kind of subjective experience comprises what we term "learning accelerants"—when students feel a slight spike in anxiety or stress.

To explore the value that students place on the tasks at hand, we used Csikszentmihalyi's descriptors of how important the task is with respect to one's future goals/plans, and to one's desire to live up to the expectations of oneself and others.[11] Finally, because we were particularly interested in the persistence of engagement, we constructed a measure of persistence—or "grit"—where we probed how determined students felt about accomplishing the task and how much they felt like giving up.[12]

We also included a set of contextual measures that describe the situation the students are in: place and time, what they were doing, what they were learning, and who they were with. Next a short set of motivation questions asked students if they were doing the activity because they wanted to or had to and whether they perceived the activity to be more like work or play—in other words, more like drudgery or an activity they looked forward to and found rewarding.[13]

Our goal was to link, in real time, students' subjective feelings with the experiences they were having with scientific practices. This would allow us to learn from the students which of these practices they found most engaging, as well as those at which they felt the most successful. We chose science activities that corresponded to the scientific practices articulated in the NGSS, including developing models, planning investigations, and using evidence to make an argument.

To examine whether PBL experiences would in fact enhance creativity, problem-solving, and exploring different points of view, we used an existing set of questions—taken from an international study—that are associated with measures for assessing potential innovative behaviors. The questions we chose asked students when, in a particular subject, they used their imaginations, solved problems that had more than one possible solution, explored different points of view on a problem or topic, and made connections with other school subjects.[14]

We assumed that instructional strategies can create environments that foster OLMs, and that this can be achieved through a PBL design principle framework—with guidelines but not scripted lesson plans so that teachers can adapt strategies to their particular cultural contexts. Other types of instructional activities may have similar effects on students' performance: for example, inquiry-based learning, where students collaborate with one another and produce a set of artifacts. In a 2000 book, Joseph Polman provides an in-depth description of his students "learning by doing" in a high-school classroom.[15] Similar in some respects to our units, his students actively work on asking questions, collaborate with one another, and produce artifacts using a variety of scientific practices. The difference between Polman's approach and ours is that we have a more formalized structure to our PBL treatment, one that was principally designed around three-dimensional learning and NGSS performance expectations. We expected that this structure would provide a clear set of boundaries regarding expected outcomes for the students and more guidance for the teachers—and, perhaps more importantly, that our methodological design, which targets effects on learning, would allow for future scale-up.

With respect to design, from fall 2015 through spring 2018 our international team conducted a series of studies in high schools located in Helsinki and in urban, suburban, and rural locations in a Midwestern U.S. state. To date, we have the responses of 1,700 students from Finland and the United States, including close to fifty thousand ESM responses on their situational daily life.[16] In addition to the ESM, we have collected (from both Finland and Michigan) student and teacher background surveys, teacher ESM responses, teacher and student interviews, and observations and videography from multiple classrooms.

Our Findings

The following studies, briefly described, provide the results of our work. Several findings have emerged that relate directly to the momentary nature of engagement and how it can be enhanced through targeted interventions. For example, our studies show that engagement is particularly sensitive to active participation in specific activities that stretch one's knowledge acquisition—such as asking questions, planning investigations, and modeling—and can reach points of saturation with some negative consequences when encouraged and supported for long periods. Each of these studies is identified by its lead principal investigator—in every case, however, international teams of senior investigators, postdoctoral fellows, and graduate students all played key roles in analyzing and interpreting the results.

VALIDATING THE IDEA OF OPTIMAL LEARNING MOMENTS

With respect to our work on validating the constructs used in creating our OLM models, the first set of studies, led by Professor Barbara Schneider and using measurements of engagement in real-time situations, shows that students had elevated levels of interest but, as predicted, the activities they were engaged in needed to be structured beyond their skill set (even though this was associated with slight increases in stress).[17] Presently, OLMs tend to occur relatively infrequently in secondary science classes—approximately 12 percent of class time in the United States and only slightly more than that in Finland. Researcher Justina Spicer and her team show that with the exception of some person-level aspirations, there are few differences in positive emotions during science classes when students were engaged in OLMs. In both the United States and Finland, those students who have aspirations to pursue science in the future were more likely to be engaged in science classes than those who reported having no future aspirations to pursue a career in science. But the magnitude of these differences is relatively small, considering how little time students in both countries reported being engaged in their science classes.[18]

This issue of how much variation can be attributed to person-level values compared to momentary experiences was recently examined by senior researcher Katja Upadyaya. She finds that there is nearly the same amount of variance between experiential moments as between people.[19] This means that there are multiple opportunities in classrooms to alter learning experiences so that individuals can react more positively, regardless of whether academic or other factors are inhibiting or encouraging their motivation and engagement. Upadyaya's work shows that learning enhancers (that is, feeling active, confident, happy, successful, and joyful) are positively related to feeling interested in what one is doing, feeling that the activity is important to oneself, and feeling like it does not have high costs; by contrast, when students were confused or bored, or feeling anxious and stressed, they were more likely to see the activity as having high costs, or as very difficult personally. If the students were presented with a situation that was very challenging, they were likely to feel uncomfortable, suggesting that, much like the accelerants, too much challenge can be a barrier to engagement.[20] For example, if students were asked to take on a task without the prerequisite knowledge or skill to be successful, they were likely to feel disengaged even if they were interested in the task.

The components of engagement in OLMs appear to have similar average effects on females and males. Two studies have shown variation between boys and girls, however, in terms of what was being taught and feelings of skill and stress. Research associate Janna Linnansaari (now Inkinen), in one of our earliest studies in Finland, found that females responded that they felt more skilled than males in life science courses, but felt below average in science courses like physics. This finding was the opposite for males.[21] Taking these results further, and including both U.S. and Finnish students, research associate Julia Moeller examined the accelerants (stress and anxiety). She found that in science classes, females tended to experience more stress in anxious situations and were more likely to report feelings of school burnout (context-specific feelings of school-related chronic fatigue, indifference toward schoolwork, and disappointment and inadequacy as a student).[22] Even though females were likely to receive higher grades than males in science classes, they paid a subjective price for it and—especially in the United States— were more likely to feel stressed.

Two newer studies, conducted by Professor Katariina Salmela-Aro, looked at engagement and burnout for both the U.S. and Finnish samples using situational data and other one-time measures on surveys.[23] Salmela-Aro has developed an instrument that is now widely used to measure the effects of burnout on students and teachers and has argued that the effects of burnout should be examined in specific educational contexts.[24] She found that some students experienced elevated engagement and feelings of exhaustion simultaneously—a major problem, especially in light of her findings that highly engaged and exhausted students are likely to develop symptoms of burnout and depressive disorders later on. She concludes that encouraging all students to become highly engaged can exhaust them, especially if the engagement is maintained over long periods of time. The major takeaway here is that encouraging engagement in momentary situations can be helpful, but students need time to regroup and recuperate. Achieving and maintaining high student engagement requires not only interesting and challenging guidance, but also recognition that this level of engagement can become debilitating if continued for long periods.

GRIT, GIVING UP, AND CHALLENGE

This notion of persistence and burnout in the context of science education is further addressed in Salmela-Aro's work on the relationship between "grit" and challenge in science classes. Essentially, Salmela-Aro and her team were interested in whether students want to "give up" in more challenging situations and if grit acts as a buffer during these times, inspiring students to persist in the task at hand.[25] Grit, defined by Angela Duckworth and colleagues as "trait-level perseverance and passion for long-term goals," encompasses one's ability to maintain interest, exert effort, and persist at tasks over long periods.[26] Duckworth and her colleagues originally conceptualized grit within personality theory, describing it as a trait—similar to conscientiousness or self-control—that relates specifically to long-term stamina with regard to goals. While grit has generally been thought of as a trait, our interest is in how it varies by situation in both the U.S. and Finnish contexts.

Grit is of particular interest in Finland. The word "grit" has often been associated with the Finnish term "sisu," which can be translated as "determination to overcome adversity," and is a hallmark of Finns' perception of their national character. The purpose of Salmela-Aro's study was to gain a more thorough understanding of the nature of young peoples' motivation and grit in different cultures, to understand how situational grit relates to challenge, and how it can be enhanced. She argues that one way to gain a clearer contextual understanding of "grit" is to relate it to challenge in specific situations. Challenges in learning situations can be viewed as the emotional or psychological costs students face while engaging in different academic tasks. These costs, however, provide students with opportunities to improve their abilities beyond what they have previously mastered, and can guide their behavior toward the mastery of new learning goals.

Salmela-Aro, using data from our 2016–2017 Finnish and U.S. samples, focused on the following items: "How gritty (determined) were you to accomplish the task you were doing?"; "Did you feel challenged by what you were doing?"; and "Did you feel like giving up?" In Finland, 8,378 ESM responses were collected from 173 students in 9 classrooms taught by 9 different teachers. In the United States, 8,273 ESM responses were collected from 244 students in 15 classes taught by 14 different teachers. Results show that U.S. and Finnish students' feelings of grit and giving up are more stable from one situation to another than are their feelings of being challenged. Only half of the variance in grit and giving up were at the student level, which means that both of these feelings are not altogether stable student traits—importantly, environment matters. The findings indicate, too, that challenge is more stable between situations in the U.S. sample than in the Finnish sample.

Overall, students' propensity to "give up" increased as the classroom activities

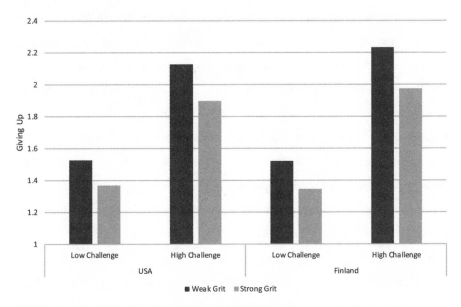

Relationships among grit, giving up, and challenge.

became more challenging—but when classroom activities were more challenging, students also reported stronger grit. Looking specifically within subgroups, students with stronger grit were 15 percent more likely to persist with a highly challenging task, while students with weaker grit were more likely to give up in similar situations. Results for U.S. and Finnish students were similar, although U.S. students were less likely to give up as the level of challenge increased.

Of particular interest is the finding that grit (or determination) varies in relationship to the level of challenge being experienced in specific situations. This suggests that "grit" is likely partially situationally dependent and can be shaped by the experiential activities at hand.

ENGAGEMENT IN TYPICAL SCIENCE LESSONS

How can teachers raise levels of student engagement in their classrooms? Graduate students Janna Inkinen and Christopher Klager examined the types of scientific activities that students were engaged in during science classes, and found that students in Finland report spending 39 percent of their time listening to the teacher lecture, the most of any activity, and the least amount of time on the computer, testing, doing laboratory work, and presenting.[27] In the United States students also report spending the most time listening to the teacher lecture (24 percent) and the least amount of time presenting, testing, computing, and calculating.

The major difference between the two groups is that Finnish students are doing a lot more listening, whereas the U.S. students report more of a balance between listening and discussing. Students report being situationally engaged for only a small percentage of the time spent in their science classrooms. When students are engaged, they report higher levels of discussing than when doing other activities. In both countries, students are less likely to be engaged when listening to the teacher or other students (these findings are similar to those of earlier work in the United States with high-school students).[28]

Taken together, these studies show that student engagement varies from situation to situation, as do the emotions associated with optimal learning moments. They also vary by gender and during certain types of activities. Challenge emerges as a more complicated idea than originally conceived in that it plays a more central role in "grit," suggesting that being determined and perseverant is highly related to the challenge or, as conceived by Salmela-Aro, as a "cost risk," and is related to feelings of stress and anxiety. The social psychological results from our work show that if one is to measure engagement it should happen situationally, but its long-term effects on a person-level are less clear. What is apparent from these studies, however, is that changing the learning environment can affect student engagement and creativity.

ENGAGEMENT DURING PBL EXPERIENCES

To assess the impact of the project-based intervention on student creativity, we employed a single-case design. Known as a reversal or ABAB design, this method establishes each classroom as its own experimental control through repeated replications of an effect.[29] In other words, during the baseline (A) phases, when teachers were engaged in "business-as-usual" instruction, we would expect to see lower levels of creativity than during the treatment (B) phases. Demonstration of the expected changes in creativity levels between baseline and treatment phases, and replication of this pattern across classrooms, allows us to make inferences about the effect of the intervention. During each of the four phases, we used the ESM, delivered via smartphones. With this approach, we are able to compare what an individual is doing and how he or she is feeling "in the moment" during traditional instruction experiences and when engaged in PBL experiences.

SOCIAL AND EMOTIONAL LEARNING DURING PBL INSTRUCTION

Graduate student Christopher Klager analyzed the relationship between "high challenge" in traditional classroom experiences and "high challenge" in PBL experiences.[30] Klager finds that during traditional classroom experiences, high challenge is associated with feelings of anxiety, stress, giving up, and confusion—results that are consistent with the average effects our researchers find when examining relationships between challenge and these other emotional states. When students are engaged in PBL activities, however, the highest level of challenge is associated with considering different points of view, using one's imagination, and feeling like the task is important to one's self and one's future. The message here is that PBL—which underscores "figuring out phenomena" by having students engage in arguments using evidence, use scientific practices, and produce artifacts to demonstrate their evolving understanding—appears to be changing the social and emotional dynamic that arises when students tackle challenging questions that, though unfamiliar, are personally meaningful.

PBL INSTRUCTION, PROBLEM-SOLVING, AND IMAGINATION

The next analysis, also led by Klager, investigated the project-based intervention for secondary chemistry and physics teachers in the United States. The mean responses on the vertical axis were calculated for the variable "exploration of different points of view in the U.S."

Additional analysis also shows that during PBL, in both U.S. and Finnish chemistry and physics classrooms, students reported a greater use of scientific practices, including the number of times they worked on developing scientific models

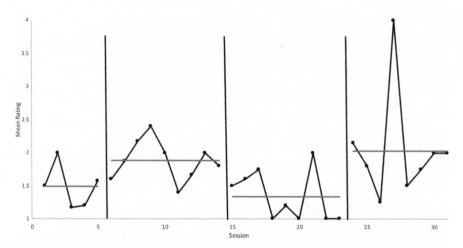

Exploration of different points of view on a problem or topic. All students in a classroom responded to the ESM survey at the same time. The vertical axis shows the average of all student ratings (on a 1 to 4 scale) for the ESM question related to exploration of different points of view on a problem or topic. The horizontal axis is the number of measurement occasions.

and on conducting investigations. They also were more likely to report using their imaginations and solving problems with multiple solutions. In the U.S. sample, the pattern is less clear during the initial PBL period, but in the second PBL instructional period we find large increases in the use of imagination and in solving problems in multiple ways; these results suggest that students and their teachers may have been becoming more familiar with the new pedagogical practices.

More recently Klager has analyzed two years of treatment data on these same outcomes for a different sample of students and teachers.[31] He finds that not only are the effects of PBL significant, but they also increase in magnitude. These results are among the most promising, because they have given us an opportunity to estimate the reliability of the effects of PBL.[32]

Finally, we asked teachers for their end-of-term grades for the students who participated in the study, as well as the students' grades for their daily lessons. With respect to daily lesson grades, we find that there is little variation on most days, with the majority of students receiving grades that reflected their full participation in lesson activities. The end-of-term grades changed only for the group that began with grades lower than the classroom average (approximately one-quarter

of our science students received "A" grades, while the students in the lowest quar-
ter tended to receive less than a "C" average, with few failures or "D" grades). We
recognize that grades are not a clear measure of learning performance, which is
why we rely on student artifacts and products as well as unit assessments (and in
the future, on a summative administrative state test). Nevertheless, we were encouraged to find that students who experienced more OLMs had higher grades by
the end of the term. We are not suggesting that these measures in and of themselves show growth in science learning—but they do mean that among the more
social and emotional outcomes we can be more definitive about our early results.[33]

The Significance of Our Results for Science Learning

For decades, policymakers in the United States, concerned about meeting national
security needs and ensuring the nation's global economic competitiveness, have
called for improvements in how students are taught science. Similar types of reports have been released by other countries in recognition of the need to revamp
science education in order to foster inventiveness, protect the environment, and
promote human well-being. The National Science Foundation recently released a
report with recommendations for cultivating STEM "innovators," with a particular
focus on groups that have traditionally been underrepresented in science careers.
The report argues that creativity and innovation are not only key for spurring entrepreneurial activity; they also are necessary for solving global problems like climate
change. The report offers several recommendations, including providing students
with real-world and hands-on activities and the chance to interact with practicing
scientists and engineers.[34]

Providing students with these types of opportunities is not new, but science
instruction has nevertheless continued to lag behind these recommendations, with
little attention given to allowing students to engage in scientific practices. Science
and engineering are creative endeavors that require that people explain phenomena and derive solutions to problems. Traditional science instruction, however,
often focuses on learning content without the context of scientific practice, offering
activities for students that leave little room for creativity. The NGSS emphasize the

integration of scientific practices into instruction, but the existence of these standards alone cannot ensure that teachers have the tools to foster creative thinking and innovation in their science classrooms. Our test of PBL, while early in its research phase, is profoundly encouraging.

When teachers consider levels of skill, challenge, and interest, student engagement levels increase and are accompanied by other positive social and emotional effects, as well as declines in feelings of boredom and confusion. We have learned that concepts such as engagement, creativity, and problem-solving are situationally specific, and share nearly equal variance when contrasted with person-level characteristics; in other words, even if a person is not interested or does not perform well in science, carefully crafted situations can alter their predilections. PBL appears to add considerable strength to the "nurture" side of learning, especially with respect to imagination, problem-solving, and taking different points of view into consideration.

PBL is not only showing early signs of positively affecting students' social, emotional, and learning experiences; it also appears to shift how teachers reflect on their own science pedagogy, which, as others have shown, helps teachers to challenge, reconsider, and evaluate notions of their own teaching techniques and student learning.[35] This type of self-reflection, in turn, tends to lead to more collaborative experiences and scientific practices taking place in their classrooms.

5. Teachers Reflect on Project-Based Learning Environments

It's a little bit difficult to put into sentences, but learning PBL has been a long process. It has been a learning process for me. It's very important that teachers are learners at the same time as their students. For me and my colleagues it's a new way of thinking. . . . It fascinates me. It underlines learning, the teacher as learner, to make science alive, to make science meaningful, and to learn what is behind science. Not only to take the results of existing science ideas but to undertake a small research path.

—AKU, a physics teacher in Finland

It was very new and it meant having to think through an entire paradigm shift that we never thought about before. It wasn't just about doing projects—we did projects before. In the beginning I wasn't quite sure what I was doing, but I knew that if I got my kids involved in it and got their ideas out, it would be meaningful to all of us. It's funny as you hear me talk because it's no longer about what I'm doing, it's about what we are doing, what me and my students are doing together. There are so many layers to it, it is so multidimensional and so exciting intellectually, really.

—ELAINE, a chemistry teacher in the United States

The process of teaching and learning science is complex and it is difficult to construct a single paradigm that encompasses it all. One of the key leaders in this field is John Bransford, who has outlined which teaching and learning processes help students create meaningful and understandable knowledge. Bransford and his colleagues underscore the importance of supporting and creating meaningful knowledge through teaching that is active, reflective, collaborative, cumulative, and contextual.[1] As he and others have suggested, these processes can be realized through activities that encourage students to use scientific practices such as conducting observations, collecting data, constructing models, and interpreting results.[2]

In the United States and Finland many of these practices, which we maintain are critical for science learning, are not used often enough. Instead, instruction in the United States and Finland tends to be teacher-led and delivered through presentations, often using PowerPoint.[3] There have been many suggestions for how to change this instructional paradigm; PBL is one of these because it emphasizes collaboration and students' active role in knowledge building. Relying on the intensive interviews we conducted with the teachers in our study (some of them several times during their units and across years), we describe in this chapter how they incorporated PBL in their classrooms and how they felt it affected their teaching practice as well as their students' learning.

Why Participate in PBL?

In earlier chapters, we explained the changing societal environments of the United States and Finland that have encouraged national reforms for how science is taught in elementary, secondary, and postsecondary schools. But as we know, national policies do not easily translate into classroom practice, especially in large countries like the United States. U.S. teachers often view national policy recommendations and mandates as both egregious and problematic with regard to managing their classroom practice and promoting student learning.[4] Finland, however, has not experienced this disconnect and acrimonious sentiment between national policies and local practice.[5] Professional teachers in Finland are involved in the planning and implementation of national education policies. In Finland, teachers are regarded as professionals because of their competence and knowledge, and because the Finnish population has a high level of social trust and regard for teachers' expertise, teachers can practice autonomously, without the need for inspection and testing. School-society-family partnerships also contribute to teacher professionalism. This respect for teachers is not widely shared in the U.S. policy community at national and local levels, nor are there many opportunities for teacher collaboration in policy-related decision-making processes. Yet despite these differences, teachers in both Finland and the United States not only were willing to learn PBL (which is consistent with both countries' new national science standards), but also viewed it as important for their own professional learning and their students' science learning.[6] Why?

Several key ideas surfaced when we asked about the teachers' motivations for experimenting with PBL.[7] The first centered on a need for a new model of teaching: one that would help their students identify and acquire evidence to respond to the important scientific questions of today. According to Phillip, a physics teacher in the United States: "I think the teaching is different. That is a big difference—there is less direct instruction. When you think about a traditional model of teaching, when I am in the front of the room giving instruction about what we are about to be doing and what we can expect to find, my students are not able to put the pieces together (part of that, I suspect, was my own teaching). But I liked the idea of a new challenge for teaching, a new model in a new way—figuring out what needs to be done differently. Now I can be saying, 'Okay, we just did this really cool experiment, how can we take what we just did and figure out what it means?'"

The teachers recognized an opportunity—one steeped in design principles and theory—to change science learning for themselves and their students. Leena, a Finnish teacher, explains, "Another main difference for me was that there is a very good theory behind everything; it is not just some new method. There is really a solid basis of theory behind all of it so the whole package was very impressive. This PBL is really something different that I have never used before. This is really something new and I always like to try something else. I have been a teacher for twenty-two years; it can become boring if one does things the same way every time."

In addition to the PBL research base, another motivation for teachers to participate was the chance to personally take part in the research experience. For example, as Aku, a physics teacher in Finland, relates, "I like the idea of a student as a researcher and the whole group (teachers and students) as researchers. PBL gives us tools for focusing on important aspects for answering questions that are useful for science learning."

Engaging in research in their classrooms was more of a fundamental change for U.S. teachers than Finnish teachers, who are required to prepare a second thesis during their pedagogical studies that involves participation in empirical research.[8] The need for and interest in science teacher involvement in research is highlighted by Elaine, a chemistry teacher in the United States: "We really need a more research-based approach for teaching science. It was a way for me to involve research projects into science learning. But my initial reflection was it was going to be very difficult. How was I going to integrate the science content and the projects simultaneously?"

For the teachers, another perceived benefit of participating in our PBL program was the opportunity to develop and provide instruction that focused on purposeful and relevant problems.[9] Teachers were quick to explain that their science instruction became more valuable when students felt they were actually learning real things instead of just memorizing textbook definitions.[10] This focus on visualizable, real-life problems helped to bring science learning more concretely into the adolescents' world. As Phillip (a U.S. physics teacher) explains, "Students need to be engaged with real-world scenarios and relevant problems. I think that there's a big shift with the NGSS, Common Core, and a lot of things that are going on today in education, away from that kind of 'baking' method of teaching, where I'm the holder of knowledge, and I'm going to give it to you."

This focus on providing meaningful and understandable problems has also been a theme of efforts to foster creativity, which includes using problem-solving skills, accepting challenges or risks, and following through or persevering with a task.[11] As our results show in Chapter 4, one of the most robust findings from our empirical work has been that PBL instruction seems to enhance students' use of their imagination, problem-solving skills, and ability to take into account different points of view.

Teacher Challenges with PBL

Change can be difficult, and PBL involves not only a new way of learning for the students, but new instructional demands on teachers. The teachers in both countries perceived PBL as challenging and were forthcoming about their feelings of uncertainty and vulnerability. Similar perceptions were voiced by two physics teachers, one from Finland and the other from the United States: "Using PBL in my classroom was challenging. The first time I went to do this I wasn't sure what my involvement would be in the lesson. I did not have everything under control. Usually I have everything under control—or almost everything—but here it was not certain how and what my role would be, but I think it went well" (Matias, physics teacher, Finland); and "Yeah, it was challenging . . . it was such a different way of teaching. I wasn't giving them information that I was asking them to regurgitate. So, what was I supposed to do? But that's a good thing, right? . . . The most difficult part, I think, was just being able to pass the reins to the students and have them take control" (Phillip, physics teacher, United States).

One area where our empirical results showed a lack of confidence on the part of the teachers was in the use of new computer technology for building models.[12] Many of the teachers had not used this type of program and expressed concern about their ability to teach it effectively. Comparable reports were also found in the students' comments.[13] As Mary (a U.S. physics teacher) states, "I struggled with constructing models in the beginning but began to realize the benefits for me and my students more towards the end of the unit. I had never used anything like this before and that made it really difficult for me." Mary further explains that, over time, she expected that this valuable activity would promote new skill development and less anxiety for both teachers and students: "I think my students didn't get as much out of it as they could have this time around, but next year they will, I think, because I will have become more familiar with the setup and what I need to do to help my students."

Use of the Driving Question

Motivated to learn how to use PBL in their instruction, the Finnish and U.S. teachers engaged in a series of professional learning activities and between-country exchanges where they observed each other's classrooms.[14] The professional learning sessions introduced the theory, allowed teachers to try out the same lessons they would be implementing with their students, and provided reading materials and lesson plans for the units. In Finland, teachers received translated versions of the overall unit descriptions and some of the lesson plans. In addition to these in-person activities, the teachers interfaced with our PBL experts in video conferences. Perhaps the biggest difference in the PBL professional learning experiences and other science reforms the teachers had tried was that now they were being exposed to new instructional tasks deeply embedded in the constructs of three-dimensional learning. Teachers were taught how to organize and deliver their units to be consistent with a "driving question," a purposeful and personally meaningful question tied to a performance expectation standard—that is, a set of scientific practices that encourage understandings of various disciplinary concepts.[15] While this focus on the driving question—which creates a need to know and motivates the learning tasks throughout the unit—was initially challenging for the teachers, they found this aspect of the PBL framework meaningful and satisfying, for both themselves and their students.

As one of our Finnish physics teachers, Aku, remarked, "The driving question is a question that has meaning and interests the students and is big enough of a question that you need many lessons to teach it. I think that it is very important to have a good driving question and that was one thing I really thought about. What was really motivating and interesting [was] at the end of every lesson, we checked to see how we were answering the question, which I found very important throughout all my units."

Another Finnish teacher, Leena, describes her initial apprehension about the process, change in impressions over time, and her willingness to accept and continue the focus on the driving question through the entire unit:

> The idea of a driving question was a new idea for me and it took some time to understand what the meaning of it was. And at first it sounded a little bit too simple: to put out one question and to get an answer for it during many, many hours. Now I think I understand a bit more, and a good driving question makes sense throughout the whole unit. For example, it could be learning about the relationship between current and water. The driving question would be how to explain motion, that would be the

starting point, how to describe motion, how to draw motion—lesson by lesson we discovered how to do it. I kept checking at the end of every lesson if we had answered the lesson questions. It was challenging; in the beginning I explained this new approach to learning and what we were going to do. I took the time because I was afraid they would not understand what we were trying to accomplish. What was amazing was how frequently they got involved in the process and it was working—and it wasn't anywhere near as problematic as I had originally thought.

The importance of the driving question became a unifying principle that gave the lessons focus and a common goal. As Richard, a U.S. physics teacher, said, "I think the driving question helps to keep the students focused on not getting too tied down on too many details—and instead look at the big picture we are trying to accomplish. It also kept my focus." The issue of focus is reiterated by Elaine, a U.S. chemistry teacher: "This approach is so focused, there is a common goal—a driving question that you and your students own from the very first day of the unit and that same goal is understood by all, is accessible to all, every day for the rest of the unit. The big piece. It really kept everything focused and tied it all together for the students and me every day."

Engagement in Scientific Practices

One of the basic precepts of PBL is that science content cannot be learned without engaging in scientific practices. In addition to asking questions, learners need to design their own experiments; plan investigations; observe, collect, and analyze data; interpret information for making claims; work collaboratively; and generate new ideas.[16] For the teachers, the most exciting part of PBL was their students' involvement in scientific practices such as planning and conducting experiments, model building, making claims, and creating artifacts to show their developing understanding.[17]

In describing his experience with PBL, Noah, a U.S. chemistry teacher, details his vision of how the scientific practices help to build critical question-askers: "PBL offers more freedom for the students to learn for themselves and actually provides students day-in and -out opportunities to perform experiments, collect data, and then reformulate what their beliefs are; . . . it also provides a lot of opportunities for them to be able to make a claim and then support it with specific evidence that they have calculated and found out for themselves. It lets them be critical thinkers, question-askers."

Planning and conducting investigations require making systematic descriptions and developing and testing theories about how the world works. Students need to figure out what types of observations and experimental designs will provide the data they need to build explanations and models that address cause-and-effect relationships. Part of this process involves choosing appropriate measures and controlling for factors—that is, variables likely to confound results. As Elaine, a U.S. chemistry teacher, explained:

> Chemistry—a lot of it is conceptual. We cannot go out and see atoms. When I get out of a pool I feel cool. Why does this happen? A lot of [the students] had misconceptions because they thought about their pores opening up. So, you can help to clarify some misconceptions. We look at these liquids and try to understand what is happening. Whatever we do has an effect; we are talking about real-life pictures. When you do something, it does go away, it goes somewhere. It may be transformed. "Your actions have reactions." The first unit allowed them to think about molecules and how they interact. The evaporating and colliding led into inter-molecular forces—why different liquids evaporated or cooled at different rates. We worked through the hard evidence they saw for themselves, the data that they collected to be able to figure out why things cool faster or slower.

Modeling helps scientists and engineers envision and explain the relationships among variables. Models can take many forms, including mathematical formulae, diagrams, and computer simulations. They represent phenomena and physical systems and processes in a consistent and logical manner. In physics and chemistry, students often deal with mathematical formulae. Students are expected to develop, revise, and use their models to describe a phenomenon. Linda, a U.S. chemistry teacher, described the experience in her class:

> We were on the unit of why salt is safe to eat, what are the elements that make it non-dangerous? [The students] had to draw out what they thought

the chemical reaction was occurring to make this happen. They came up with the best questions as they were modeling—thinking about what's going on, making connections with previous units.

Phillip also found modeling a valuable challenge for his students (and himself):

Model construction was something new for me and my students. There was a big focus around constructing and revising models. The value of that is it allows students to take an abstract thought or concept and put it on paper and try to simplify it. It is really a meta-cognitive process that they are thinking about—"Okay, here is what I just learned." Now, how could you put it in its most simplified terms?

Making claims about something they just learned requires students to gather and use evidence to justify their ideas about how natural phenomena occur. Science is all about making an argument for why one explanation is better than another. Through the process of making claims, students learn what it takes to be a critical consumer of science, one who can weigh differing points of view and judge the quality of a scientifically reasoned argument. Richard, a U.S. physics teacher, puts it this way:

I like them being able to use evidence to be able to make a claim. It's been rewarding to watch them take the small steps toward learning, I guess toward understanding the phenomena. I guess in the end, even though it seems like we're moving slow, sometimes when they do learn it this way it seems as though they've really got it. So, I enjoy that, that's rewarding to me. They can thoroughly explain some concept or some phenomena and you can tell it's not just spitting something back to me and they can begin to ask even better questions to go on at a higher level.

Creating artifacts is a key component of learning science. By creating artifacts, students reconstruct their developing understanding of phenomena as they actively manipulate science ideas. As students explain phenomena, they extend their understanding far beyond linear, discrete information to connect ideas across a variety of concepts. This process also allows both teachers and students to critique and provide feedback on ideas and explanations.

Through experiments and building models, the students produce artifacts— for example, they design and build electric motors that they are able to start, then photograph and take videos of them running for further analysis. As U.S. physics teacher Richard explains,

> The students built these Maglev cars that went down a Maglev track, and then later they built little motors that had to pick up a load that had to work—the idea was to show some of the interactions between magnets or electromagnets or magnets for the motor. By building the motor, I was able to tell by their struggles and even [by] which things didn't work what they were understanding, and what they had done to try and make their motors work. By looking at all the motors, even the ones not running over the course of the unit, I was able to tell if they were grasping the main idea of energy flow in their artifact. By the time they were building the motors, they were interacting better as a class, working together to solve a problem.

The process for evaluating artifacts—physical products—has been a challenge for our team, especially when teachers use them as evidence of students' three-dimensional learning. To solve this problem, we worked with our teachers to develop a proof-of-concept study, where we created a scoring rubric and applied it to artifacts from three classrooms. Preliminary results show that slightly more than half of the students' artifacts were scored as using scientific practices, driving questions, and crosscutting concepts. This early-stage work suggests that artifacts can be scored for measuring students' knowledge of three-dimensional learning, which in turn can help refine artifact assessments to be better aligned with NGSS recommendations for three-dimensional learning.[18]

Collaborative Learning Experiences

As stated in the National Research Council's *Framework for K–12 Science Education,* "Science is fundamentally a social enterprise, and scientific knowledge advances through collaboration and in the context of a social system with well-developed norms. Individual scientists may do much of their work independently or they may collaborate closely with colleagues."[19] Successful learners share, use, and debate ideas, creating a community that supports making connections between ideas. This focus on collaboration was viewed positively by all of the teachers in both the

United States and Finland. Phillip, the U.S. physics teacher quoted earlier, shared: "I think collaborative science is of utmost importance for the students. It is just kind of my ongoing development of what I think it is going to take to get kids to be successful. As Deborah Peek-Brown [one of our PBL experts] mentioned, 'Students need to get out of their science curriculum and traditional science education and learn how to work with groups of people that they're not necessarily friends with or that might disagree with or are different from them.' Getting those diverse perspectives and working through problems together is critical in the twenty-first century and for what our kids will face in the future."

Mary, also a physics teacher in the United States, agreed. "It had them working together—that was a big thing—my students do not like to work together. But they found out when they were able to help each other they felt better about themselves and each other. When a lot of the kids had the same questions and were all struggling at the same time, having them work together turned out to be a reassuring experience for many of them—meaning that it was okay not to understand things completely, and that they could work together to figure it out."

In Finland, where students typically do not collaborate as much as they do in other countries, Matias, a physics teacher, described how the collaborative activities, supported with the PBL framework, helped students discuss and work together on a problem:

> It was really nice to see how they talked about the phenomenon and together they discussed about the answers. It wasn't easy for them in the beginning and I always had to push and drive them to talk to each other and discuss the things they did. It enhanced skills to collaborate with each other. In a normal lesson they do not collaborate as much. This is one of the best things about [PBL] collaboration. We usually build a galvanic cell together. This time they built it in small groups and had to answer the question, "Why do you get a reaction?" "What is the volt measuring?" They had to take many steps for several days before getting the answer to this question. Usually I control the situation more. I always tell them what to do and we make the conclusions together. I think it worked well in the past, but now they had to make their own conclusions and the end result was maybe the same or maybe even better because more students were mentally involved in the process of learning. They were working together.

Reflections on the Challenges of Teaching PBL

Students have varied interests, skills, and knowledge, and these differences were particularly evident in the process of "figuring things out." Even though students collaborated with each other, PBL did not always have a similar social and emotional effect on all of the students. Some students appeared to "take to it" more easily than others, while others (especially high-performing students) were more resistant. Phillip explained:

> Not all students figure it out. I cannot say that every student figured everything out, every day, right? That's going to be the case no matter what. But, with these project-based units and lessons, the students that were likely to figure it out were my students who struggled with more traditional forms of teaching and learning. In my class there were students who were really engaged and able to approach a problem, think about it in an abstract way, from multiple perspectives, and come to a solution; and if they were not able to come to a solution, they were able to walk away from it saying, "Okay, what did we do? How could we have come to a solution? What did we miss?" Whereas my high achievers would get really frustrated because it was not a logical sequence of steps to follow to get to this clear-cut answer. It was like, "You're asking me to think, you're asking me to think about how I am thinking. You're asking me to ask questions—you should be the one asking questions."

A similar perspective was articulated by Finnish physics teacher Leena: "In our school we have quite talented students so I think they understood what the next step was and what they were expected to do, so following the PBL framework for them was not difficult at all. But these same talented students were the ones that just wanted to be told the answer. Some students said, 'Why don't you just tell us? Why do we have to figure it by ourselves?'"

For the struggling student, the instructional change was seen as academically supportive and encouraging. Elaine, a chemistry teacher in the United States, reveals, "Honestly, I had one student who stopped me at the beginning of the school year and she said, 'I never understood science,' and she was so happy that she actually got something out of science class last year. She never felt that way about science before. For the kids that struggle I think there is a sense of safety. Every person has something to contribute that might add to others' models and others' ideas."

Over time, there seemed to be an emerging consensus among both high-performing and struggling students that PBL did not lower their grades (in fact, it actually improved the grades of struggling students the most) and that it was a satisfying method for learning.[20] According to Finnish physics teacher Aku, "The most successful learners, in many cases they want to study so that they pass the tests and get very good marks. PBL learning—it is a little bit more complicated than the conservative way of teaching and learning. They had some problems studying, they were a little worried, but in the end they have found PBL a nice way to study and have gotten good marks afterwards."

In a U.S. school with attendance and disciplinary issues, both the successful and struggling students viewed PBL positively. As Mary, one of our U.S. physics teachers, explained, "The ones that are more attentive, the ones in class the majority of the time, where absence is not an issue and they generally do well in other classes too—it had meaning to them, it was motivating for them to do well. And even if they don't do well in their other classes or this one, they asked a lot of questions and did better."

Both teachers and students felt that PBL supported science learning because it contextualized content, pushed students and their teachers to collaborate with one another, and fostered a more active learning process that involved using big ideas and scientific practices to figure out phenomena. Through this process teachers and students were able to plan, engage, and reflect on their learning.

TEACHERS' PERSONAL REFLECTIONS ON PBL

As we've seen, many of the comments by teachers from both countries were positive. Nearly all the teachers articulated a common understanding of how using PBL influenced their classroom experiences. In particular, on having the student drive the work, Leena, a Finnish physics teacher, related: "There was a difference in how I teach. When I teach, I talk. Now I really had to try to be quiet and let them do all the work." On asking meaningful questions, Aku, also a physics teacher from Finland, shared: "It has been a learning process for me and my colleagues—it is a new way of thinking and teaches wholeness, how to teach science as a whole system. It underlines learning, teachers as learners, it makes science alive and meaningful and [focused on] what's behind the science." And on practicing science as a collaborative and iterative activity, Phillip explained:

> I think the opportunities for PBL require revision—that is pretty unique, especially because of the stress placed on the idea that it is not the final product that we are doing today. It is a constant revision and we all look at it and continue to improve. And then there are the times beyond when I am teaching science that I find myself using these same features. Making me more apt to be more engaged, share ideas, and be an active listener— not always ready to jump right in—like teachers often do. I find myself making sure that every voice is heard and recognized as important.

Some Concerns and Critiques

This was an initial attempt to implement PBL in high-school physics and chemistry classrooms with teachers who were new to it. As this was a pilot study, there were problems. Some of the experiments did not work, and some of the materials did not do what they were supposed to do. Some of the teachers felt they needed more support, especially if there were management issues in their classrooms. Several of the teachers thought the lesson units were too short; others thought they were too long. Some found the grading process challenging, especially for the activities that "produced group artifacts—that had the group (rather than an individual) take ownership for the projects." And finally, some teachers worried about explaining to the parents what their children's contribution was and how that translated to an individual grade.

We are not trying to present PBL as a panacea or to suggest that everything worked perfectly. It did not. Teachers did not hold back in their comments, and were candid in exposing their own vulnerabilities and the uncertainty of trying something different—a level of honesty that was reflected in all of the types of data we collected and analyzed. What is perhaps most surprising, however, is how the quantitative data reflected some of the teachers' impressions and how those impressions mapped onto the students' responses.

A Look at the Teacher ESM Data

All of the teachers participated in the ESM, receiving a shorter version than that given to the students (see Appendix C). The teachers also answered a background survey that included some items from the Teaching and Learning International Survey (TALIS) and Carol Dweck's growth mindset questions.[21]

Similar to the students in our study, when teachers experienced optimal learning moments (OLMs) they were more likely to have enhanced positive feelings, such as feeling excited, confident, proud, and happy. They were also less likely to experience detractors—feelings of boredom or confusion—but experienced a slight elevation in their accelerants, feeling increased stress and anxiety. As with the students, OLMs tended to happen infrequently. During science classes, the average proportion of time teachers reported experiencing OLMs was about 27 percent; for students the percentage was a bit lower, with OLMs occurring about 19 percent of the time.[22]

We wanted to know if students and their teachers were more engaged in science when involved in specific learning situations—in this case, when involved in PBL activities. To assess this, we examined whether the teachers' challenge, interest, and skill—three important components of OLMs—were elevated above their average during PBL activities. Analyzing the ESM data from our teachers, we found that during PBL, teachers were significantly more likely to feel challenged and interested in the activities at hand. Although not statistically significant, teachers also appeared to feel slightly less skilled when engaging in PBL instruction. This is important, because it corresponds to comments made by the teachers who, as noted earlier, reported that teaching the PBL units was of considerable interest but that they also felt challenged when working in these environments. The fact that they felt less skilled makes sense—for many of the teachers, this was the first time that they were trying out the units in their classrooms.

With respect to scientific practices, some interesting patterns emerge for both the teachers and the students. Both teachers and students rated planning an investigation as a highly positive OLM experience and communicating information as one of the least positive OLM experiences. The patterns diverged on the next two highest and lowest scientific practices, however. Teachers rated developing models and making arguments using evidence as highly positive OLM experiences, and gave their lowest ratings to defining problems and designing solutions. These lower-rated tasks were likely related to the teachers' familiarity with them, because they are used in traditional science teaching. The students' highest-rated experiences tended to be more active: specifically, defining the problem and designing solutions, the opposite of the teacher ratings. Here the students took on the role of the teachers, asking the questions and designing their own solutions. The experiences the students rated as of low interest were developing models and communicating information—skills that, while critical for making claims, were ones the students already could do somewhat proficiently. They viewed these tasks with lower interest—probably because they offered less of a sense of ownership and the students felt only moderately challenged while doing them.[23]

These ESM results are preliminary and subject to all of the vagaries of data analysis, including missing data, lack of clarity with regard to what is being asked, and a multitude of concerns over how to measure responses.[24] But even with treatment effects, and in contrast to other social and emotional measures found in retrospective and one-time measurements, the trends, patterns, and significance measures remain reliable and valid.

In addition to the ESM surveys, we gave the teachers a background questionnaire, the results of which show that our sample group had a high growth mindset—that is, they are teachers who believe intelligence is malleable and can be changed, who support the idea that thinking and reasoning skills are more important than acquiring specific content, and who prefer having their students solve problems themselves. When teaching, our teachers with this kind of mindset reported feeling enthusiastic, strong, and as if time was flying by. Our teachers generally did not report indicators of burnout or dissatisfaction with their job (a finding that is at odds with the recently released profile of teachers in the U.S. *Schools and Staffing Survey*).[25] As we have discussed, these problems (that is, burnout and dissatisfaction) are not as pressing an issue for teachers in Finland as they are in other countries.[26]

These results are indeed promising, and we are now oversubscribed, with a number of school districts in Michigan and other states wanting to become part of the work. We are also receiving a large amount of interest from countries other than the United States and Finland. We are committed to conducting a larger efficacy trial with more students and teachers using classic randomization procedures.[27] But we will also be making our materials, including assessment items, open source in the near future.[28]

One of the most important conditions of successful education reform is the commitment and involvement of the teachers. Our teachers have been willing to experiment even when challenged and unsure about the final result. In many ways, what we have learned mirrors what was found in the pathbreaking book *Trust in Schools*.[29] Efforts at change require the melding of well-conceived initiatives—PBL being one—that are undertaken as cooperative activities by all of the participants. Our teachers viewed their involvement and engagement in this work as not just another opportunity to do projects, but also a chance to learn from a well-designed, theoretically rich initiative that is recognized by experts as both valuable and timely.

PART III

Active Engagement

A PATHWAY TO SCIENCE LEARNING

6. Encouraging Three-Dimensional Learning

Our results implementing project-based learning with over 1,400 students and fifty teachers in twenty schools in the United States and Finland have been promising, showing that student engagement and learning can be enhanced and that teachers can successfully change their instructional practices. What has not been addressed is how the broader educational system can move some of our ideas forward and affect the science environments of many more students and teachers. Or, for that matter, how families can become more actively involved in providing meaningful science experiences for their children and themselves.

Our vision is not to make everyone a scientist, but to reach some students who may be considering a career in the physical sciences and technology, especially women and underrepresented groups—and further, at a personal and societal level, to increase science literacy so that this and the next generation can enrich and sustain life on our planet.

Study of Engagement

One of the most important messages of our research is that engagement is situational. It is not an omnibus concept that applies to all situations or that can be sustained in all circumstances. Even for the most enthusiastic students, being in an extremely challenging situation without the appropriate skill set or listening to

a teacher talk about science without the opportunity to discover scientific principles for oneself could be disengaging.

Engagement is clearly essential to learning, but what is often overlooked is
how it is related to other social and emotional factors that are critical for learning
science.[1] In many science classrooms, instruction is typically not focused on investigating a purposeful meaningful question. Instead, teachers often try to enhance interest by connecting students to scientists, or by visiting a museum or an
environmental center. But it is not enough to simply show other people doing science; learning science has to be internally meaningful. This is why students have
to be involved in the actual "doing" of science.

The science education communities in the United States and Finland recognize that to increase science learning, students need to make sense of phenomena
or find solutions to problems that have personal meaning for them—problems
that require "figuring it out" through using disciplinary core ideas, scientific practices, and crosscutting concepts. They also know that if the phenomena being explored are too complex and the students do not have the skills needed to plan a
reasonable strategy, they can easily become confused, causing their interest and
persistence to drop. If the question is obvious or has been taught multiple times,
asking students to solve it turns the discovery process into one of drudgery and
boredom.

Taking seriously the recommendations of the *Framework for K–12 Science Education* and the Finnish Curriculum aims, we realized that to enhance engagement
in science the first goal should be to change instructional practices from passive to
active learning—which is consistent with research from the learning sciences that
shows how people learn.[2] Fostering engagement by posing challenging questions
of personal interest to students—for which they have many but not all of the background skills and knowledge to solve, so there is room for increased learning—is
key to experiencing optimal learning moments. During these OLMs, students are
likely to have heightened feelings of success and confidence—all of which we assumed would lead to academic learning and social and emotional development.

We show here the OLM model and its relationship to multiple outcomes for

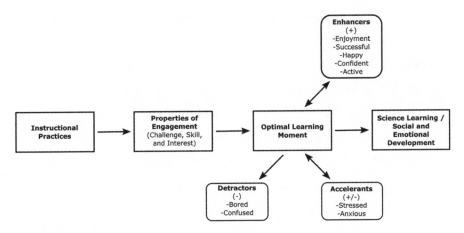

A model of optimal learning moments.

learning.[3] Essentially, the model begins with instructional practices that create an environment in which science learning and social and emotional development can be enhanced.

In this model, the properties of situational engagement are challenge, skill, and interest. Interest is not merely a personal predilection but can also be an authentic motivator for the student when he or she is presented with a problem that is relevant and meaningful. Consider the driving question in the PBL Forces and Motion unit, "How can I design a vehicle to be safer for a passenger during a collision?," which is likely to interest the many high-school chemistry and physics students in the United States who have recently learned or are learning to drive. The challenge component energizes and guides the learner toward explaining a phenomenon or solving a problem, neither of which can be achieved without engaging in a particular set of tasks that are designed to build on existing skills and encourage students to use their imaginations. Knowledge and skills here are viewed as incremental and domain-specific, yet students acquire the necessary skills to complete the unit by becoming actively involved in behaviors such as designing and building a model or acquiring and interpreting evidence to justify claims. These types of behavioral skills are likely to have transference to other cognitive tasks—in contrast to the memorization of terms and equations, which are merely ends in and of themselves.

When students are in OLMs, there are other subjective experiences occurring at the same time that can enhance or deter learning. First are the detractors—feelings of boredom if the tasks are not challenging enough, or instances of confusion if the skills needed for the tasks are too complex. Second are the accelerants—feelings of anxiety or stress, which often occur when the steps toward solving a problem have not been carefully scaffolded, or laid out sequentially over time. If too much pressure occurs, students can become disengaged or paralyzed.[4] In moderation, however, anxiety can be beneficial by activating learning, for instance by stimulating interest and inspiring a student to seek out solutions to a challenge.

Finally, there are the enhancers—the positive affective emotions that are likely to be improved when in OLMs. These include enjoying what one is doing and feeling happy, successful, or confident. These subjective feelings, we suspect, can help sustain engagement in challenging tasks, but when there is not enough progress toward completing them, or little encouragement or support is received, these feelings won't be enough. Enhancers, much like other social and emotional feelings, have boundaries that are situationally specific.

If challenge, skill, and interest are critical for engagement, which instructional strategies are most effective in creating environments where academic, social, and emotional learning can occur? Here is where we turned to PBL, with its design-based principles and innovative curriculum development, professional learning, and assessment techniques that we predicted would enhance learning. The design principles of PBL are particularly important, because they lay out clear criteria for setting up a unit (which will be centered on figuring out a phenomenon or solving a problem); they articulate observable actions (by incorporating and requiring the use of disciplinary core ideas, scientific practices, and crosscutting concepts specified at the lesson level); and they offer reliable metrics for identifying the fidelity of implementation of PBL in multiple situations. PBL also involves ongoing professional learning, whereby teachers receive support as they adopt PBL strategies in their classrooms. Finally, it allows for the use of various assessment techniques and, most promising, for students to show models and give explanations of their work.[5]

Recommendation 1: To enhance situational science engagement, environments should be crafted that provide instructional strategies—in this case

PBL—in which experiences are created that promote opportunities for personally meaningful interest, skill improvement, and challenging ideas that can be linked to scientific, academic, social, and emotional learning.

Engagement and PBL

There are several other reasons that we selected PBL to test our model of engagement. PBL's design principles, developed decades earlier, reflect the same practices that are emphasized in the K–12 Framework, NGSS, and recent Finnish curricular reforms. The NGSS did not specify a particular curriculum, so we adopted PBL because it had a track record of creating curricular units that exemplified similar aims. By creating new units, we were able to further test PBL's design principles. We also began to fill an important need to improve the academic learning and social and emotional development of secondary-school students taking chemistry and physics, courses that are often gatekeepers for entrance into postsecondary science programs.

One of the important innovative aspects of PBL is its focus on social and emotional learning, on lessons that are participatory and collaborative. Young people can become engaged when presented with questions that are meaningful to them and can be solved collaboratively using scientific practices and by learning and debating ideas. When students are engaged in working on an experiment or model in a systematic way, with a real problem that is solved through carefully scaffolded activities that result in a tangible artifact, they are more likely to enjoy the experience and feel confident about it. Accomplishing tasks collaboratively with classmates not only emphasizes how science typically progresses today, but also provides opportunities for peer interactions among students with different skill sets and demographic characteristics.

But PBL is not only an instructional intervention; it is also a system of reform, exhibiting several of the components highlighted in the book *Learning to Improve* by Anthony Bryk, Louis Gomez, Alicia Grunow, and Paul LeMahieu.[6] Their model of how to bring new ideas for improvement quickly to schools encapsulates several of our "system model" ideas for advancing science learning. In particular, our model of science learning aligns with four identifiable precepts of improvement science that Bryk and his colleagues describe.

First, the goal of our work is to improve the core experiences of teachers and students in science. We did so by designing a system that involved the teachers from the onset, with professional development learning experiences that were first in-person and then continued with virtual support while the new PBL units were being taught. The teachers were and remain integral players in the process of re-form, as we iteratively modify our units and rubrics for scoring the formative as-sessments and products. Many have even volunteered to help their colleagues implement PBL in their own classrooms. Throughout our field test, several of the teachers in the United States and Finland who initially assumed this leadership role have continued to work with us on refining our system of learning science.

Second, our system model focuses on how to change daily instruction in sci-ence classrooms (with the consensual agreement of school and district leader-ship).[7] It provides clear directions on how to move from teacher-led instruction to having teachers and students solve problems and figure out phenomena. It is framed in language that is specific to the concepts being taught and embedded in routines not unfamiliar to teachers and students, but it also underscores the im-portance of students' agency and participation, and of having them use scientific practices much like professional scientists and engineers do.

Third, we were very cognizant of the organizational dissimilarities between our schools and the cultural histories and values of their districts, and most im-portantly, how these distinctions applied to our two countries. Although we were a cross-national team working collaboratively on ideas, research approaches, and instruments, we were aware of our cultural differences at both the societal and classroom level. At the societal level, the lessons were uniquely designed to match the cultural and curricular goals of each country. At the classroom level, our intent was never to create scripts of instruction that would be followed exactly by teachers in either country, but rather to offer suggestions for changing the routines of in-struction to better exemplify three-dimensional learning and the principles of PBL.

And fourth, we have developed an assessment system that is formative. We have worked extensively on developing new forms of assessment tools that capture whether students have achieved three-dimensional learning—mastering disciplinary core ideas, scientific practices, and crosscutting concepts. Our PBL units also involve technology, especially in modeling tasks, but also as part of our assessment system. And because our larger global society is concerned with objective evidence regarding the outcomes of curricular reform, we also use standardized testing protocols designed by independent sources. This is not to say that we ignore formative assessment tasks in the PBL units—we rely on them, and have observation protocols and procedures that verify their implementation and use. In fact, our assessment tasks are especially appropriate for gauging students' learning, and thus for optimizing learning environments. But until there is wider acceptance by policymakers, school administrators, and parent stakeholders to alter their expectations of how to measure performance, we will be employing independent evidence of our results—in the form of standardized summative tests. As we have expressed previously, we expect that our instructional strategies will encourage students to successfully tackle science problems that transcend the science material often covered by traditional curricula.

> Recommendation 2: To activate science system reform, we need to develop a plan for improvement built on design principles that are meaningful to those engaged in the process, that have a track record based on real-life experiences, and that can be measured reliably.[8]

Professional Learning and Teacher Education

The use of scientific practices, a cornerstone of three-dimensional learning, has important implications for teacher education, especially within the United States. Scientists study the natural world and propose explanations based on the evidence they find. Engineers study the design world and propose design solutions to problems based on testing. These skills require that students carefully and thoughtfully reflect on and interpret what they observe, which requires planning, monitoring, and evaluation. Teachers often do not have the requisite knowledge or experiences to develop and implement these types of activities in their classes. Many of the practices required in the PBL units are new to teachers, whose typical instruction methods are teacher-led and rely on scripted material. Changing to a different model of instruction can be challenging—and indeed, the U.S. and Finnish teachers in our study did report feeling challenged initially but became more comfortable and positive over time.

One of the most significant findings in our collaboration with Finland has been learning how Finnish teachers are trained, which is quite different than in the United States. As mentioned earlier, empirical research studies—a central part of the Finnish teacher education programs—are not part of most standard U.S. teacher preparation programs. Many U.S. teacher education programs do touch on issues related to measurement, analysis, interpretation, and evaluation, but these commonly understood concepts are taken from books and conveyed through lectures and discussion, not actual practice. How are secondary school teachers going to motivate their students, be their role models, and assess their students' performance if their only exposure to practice is passive? How will they learn to teach in a way that encourages problem-solving and other more active tasks that are a part of doing science?

We simply cannot expect teachers to employ scientific practices if, in fact, they have had limited experience using these tools in their undergraduate training or in their professional learning experiences. There are several lessons we have learned about teacher education in Finland that could assist the United States in better preparing teachers for reforms in science learning. The first important consideration is that, in Finland, a teaching degree is a master's-level program with three stages: first, students are expected to master a specific science topic; second, they are expected to take a series of courses that identify how to teach that specific topic; and third, they take another series of required courses on pedagogy, human development (including social and emotional development), and technology. Students are also required as part of the program to conduct an original empirical study that reflects many of the scientific practices described in the K–12 Framework.

In the United States, pre-service teacher education and alternative certification programs often do not include opportunities to engage in empirical scientific investigations where the tools of scientists are actually used. This is indeed problematic, because it places teachers in a somewhat vulnerable position as states and districts adopt or adapt the NGSS recommendations, which expect students to use scientific practice, disciplinary core ideas, and crosscutting concepts to explain phenomena and solve problems. PBL offers a chance for science teachers to engage in three-dimensional learning practices and use technology with their students. The activities, tasks, and assessments that PBL offers help teachers prepare their students for learning situations and problem situations that they are likely to encounter in their future education and adult lives.

Given the recommendations of the K–12 Framework and the NGSS, it is time that U.S. educational policymakers seriously consider how to reintroduce scientific practices into teacher training programs—like in Finland, where aspiring professionals are encouraged to actually conduct their own empirical investigations of a problem and defend a thesis or an argumentative understanding of a phenomena. This can be observational, whereby aspiring teachers ask questions that can be answered only if they use big ideas to figure out the phenomena, make observations, make assumptions about patterns, and develop and test for possible explanations. This could also include modeling, whereby potential teachers construct and use a model to communicate their understanding of how something works. The idea here is to use models as tools that can support thinking about phenomena and to consider alternatives when constructing a model and interpreting results.

Our research is not a critique of the U.S. teacher education experience, but through our multiple visits to Finland and working with our Finnish colleagues, it has become clear that we must advocate for an undergraduate teacher education that gives students a deep understanding of teacher-related practices, big ideas, and the practices of actual scientists and engineers. We recognize that changing U.S. teacher education programs—which exist in diverse colleges, universities, and alternative independent programs such as Teach for America—will be complicated. But we feel compelled to share what our rich and deep collaboration with Finland has shown us about how the United States can improve its approach to teacher education.

Finland, even with its highly praised teacher education program, most recently decided that its teacher pre-service program and in-service professional development experiences needed to be reformed. This decision was based on recent shifts in demographics and the persistent demand for lifelong learners with capacities for certain skills, including but not limited to solving new problems, making reasonable inferences, and using emerging technologies. The newest Finnish National Reform effort emphasizes the importance of using twenty-first-century competencies, technology, and collaboration with other teachers to develop classroom practices, taking into consideration variations in different school contexts. The reform effort's report outlines what should happen not only in the pre-service program but also in continuous lifelong professional development.[9] It is interesting to note which new skills Finnish educators and scientists identify as the knowledge base for teachers: first, subject matter knowledge; second, interaction skills for collaboration, planning, implementing, and assessing student performance in their own classes; third, digital skills; and fourth, research skills that are necessary for consuming research-based knowledge, reflecting on personal pedagogical views, and developing strong networks with students, parents, teacher experts, and content experts. This is clearly a list worthy of consideration, commensurate with those of others who have called for reforms in U.S. teacher education.[10]

Most recently, over 32 million dollars have been allocated to this effort in Finland, with different universities proposing to conduct pilot studies to achieve these aims. Finland is determined to ensure that its students are equipped with skills and knowledge for future work—work that is likely to be quite different from today. Along with the acquisition of skills, this new Finnish reform emphasizes collaboration and supports the building of networks for understanding the diversity of school cultures and for designing assessment techniques that take these differences into consideration.

It is sobering to think that Finland, whose teachers already receive advanced-degree pedagogical training, has embarked on a new effort to reform its teacher education program because its policymakers see its current program as falling short of many of the conclusions identified in volumes 1 and 2 of the U.S. National Academies report *How People Learn*. This report concludes that all learners share basic cognitive structures and processes that develop throughout the lifespan and are shaped interactively by context and cultural influences. Because the brain can retain knowledge and use it flexibly in making inferences and solving new problems, it is important to help learners acquire mental models that can facilitate new learning and, with conscious intervention, alter biases that may have formed from prior ways of thinking.

The takeaway here for the United States is not only the importance of redesigning pre-service education but also the value of continuous professional learning for teachers, so they and their students can acquire skill sets to help them throughout their lives. Our model considers professional learning to be an essential component of advancing science teaching and learning. The additional emotional and collegial support that teachers receive through professional learning from the PBL teams helps to build a sense of trust—not only with the developers, but with other colleagues involved in learning new instructional strategies as well.[11] Conferring, collaborating, and sharing instructional experiences are all part of our intentionally coherent design for professional learning—practices that we argue are fundamental for the successful implementation of science reform.[12]

> Recommendation 3: Teacher education should target aspiring and practicing teachers with opportunities not only to acquire empirically driven scientific practices such as those identified in the K–12 Framework, NGSS, and Finnish Core Curriculum, but also to expand cultural awareness, problem-solving, and technological skills, and collaborate with colleagues and science education researchers.

Strengthening Research Quality through Collaboration

We are cognizant of the changing landscape of research, and even before the European Commission issued new standards for protecting confidentiality, registering studies, and managing and archiving data, we decided to replicate in Finland all of the practices we currently use in the United States, including the storage of our data.[13] Combining our U.S. and Finnish datasets, we have made our data accessible for verification, reuse, curation, and preservation, a process that we executed through the Inter-university Consortium for Political and Social Research (ICPSR). These research practices are fundamental for ensuring that high-quality scholarship is undertaken across our countries.

In addition to adhering to high standards for our empirical work in both Finland and the United States, we have provided collaborative opportunities for young scientists to share experiences across countries—and in the process, develop new networks and receive helpful critiques on their work from diverse audiences and global experts. In keeping with our values of transparency and open access, we continue to share our research findings and scientific outcomes among researchers prior to publication in international scientific meetings, seminars, and workshops.

We believe that this step in reaffirming high-quality research practices has been an exemplar for our students and colleagues—including our teachers, who have become active participants in our research efforts, developing curricula, collecting data, and training other professionals in the conduct of our work. As social scientists, we have had the opportunity to reflect on our assumptions and practices and learn from each other. One of the most critical and impressive parts of our collaboration was the gradual and deepening understanding of Finnish culture—which we quickly understood would be difficult to replicate in the United States.

Finland has suffered through major famines, unprecedented immigration, and foreign invasion. As a consequence, it achieved its independence only about a hundred years ago. Many leaders of the Finnish revolution were teachers and viewed as heroes. These early teacher leaders became identified with the importance of learning for autonomous self-reflective choice. With limited natural resources, Finland's major resource is its population, its human capital that has survived these conditions and faced challenges with "sisu"—a determination and persistence that defines its distinctive national identity.

Finnish people, according to recent international surveys, are the happiest population in the world—but they too are facing some of the same problems as other countries. Not unlike other countries that have taken in a diverse immigrant population, Finland is now facing the task of enculturating a growing populace with dissimilar values and skills among their students and employable workforce, as well as increasing demands on its healthcare and other societal services. And while the country as a whole holds teachers in high regard and trusts them to provide all children with an excellent education, the teachers themselves are increasingly showing signs of burnout—which in Finland shows up as increased stress, absenteeism, and feelings of inability to work.

Finland has a very different strategy, however, for dealing with its education problems—one that has important implications for the United States and other countries. When a problem arises, particularly within the education system—such as the need to increase innovative entrepreneurial activity or reform science practices—it is viewed as everyone's issue. The government calls on everyone to participate in proposing reasonable solutions and reaches out to the teachers, bringing them into the reform process and creating alternative solutions that the teachers are willing to experiment with. This is perhaps best exemplified by how Finnish teachers have willingly adjusted their science standards to reflect a more practice-oriented type of instructional strategy, or have held national competitions to stimulate technological innovations that can enhance students' learning.

One area where the United States has been especially helpful to Finland—especially as its school-age population is becoming more economically and culturally diverse—is in emphasizing the importance of extending educational opportunities to all students, including those from diverse backgrounds and recent immigrants. Appendix D of the NGSS highlights effective strategies for economically disadvantaged students, students from the main minority racial and ethnic groups, students with special needs, students with limited English proficiency, students of both sexes, gifted and talented students, and those in alternative programs. In trying to achieve equitable learning opportunities for all students, the NGSS recommends that teachers value and respect the cultural differences of all students; incorporate students' cultural and/or linguistic knowledge with disciplinary knowledge; and make available adequate school resources that support student learning.[14] We support the NGSS recommendations and have deliberately created PBL learning experiences designed to involve all students in multiple experiences, recognizing their diverse backgrounds.

We live on an increasingly smaller globe, where we need to learn from each other. PBL offered the mechanism by which we could engage with each other in a substantive research project that in many ways mirrored the concepts emphasized in the K–12 Framework, NGSS, and the Finnish curriculum. From our Finnish neighbors we learned more about the professionalism of teachers and the need to preserve their autonomy and recognize their expertise. The U.S. scholars brought to the table the original design components of PBL. Together we have reinvented what can be accomplished with science learning reforms. Our work is collaborative, cooperative, and shared, which is why this book is the product of work by our four investigators as well as our students, postdoctoral fellows, and collaborating teachers. Although they have a stronger emphasis on collaborative curriculum development, the participating Finnish PBL teachers also reported finding these experiences useful for honing their skills and sharing their experiences with other teachers. Through these efforts, in tandem with the study team's presentations at multiple scientific professional meetings, both nationally and internationally, a new PBL learning community is emerging in the United States and Finland and among other countries interested in the adoption of PBL units.[15]

> Recommendation 4: To advance high-quality scientific scholarship in science education, which can have a global reach, we need to strengthen and build bridges with other international scholars, to explore important questions using rigorous methods, and to make policy-based evidential claims.

Building a Personal and Societal Commitment to Science

Learning science is a societal issue and it is time to think about scientific literacy not only in the context of chemistry and physics, but also in terms of science literacy— that is, as not just a body of knowledge in a particular area, but rather as a shared belief in the value of science for knowing our world. We also need to understand how scientists work by engaging in the "doing" of science practices.

For too long we have ignored science literacy, but we are learning quickly that technology, whether we interact with it by driving autonomous cars or buying a seat on a spaceship to Mars, is becoming an ever greater part of our daily experience. With these dramatically new ways of living, what does it mean for people to try to understand the complex changing world we live in? The National Academies of Sciences, Engineering, and Medicine recently issued a new report on what science literacy means for today.[16] They identified aspects of literacy including content and epistemic knowledge; foundational literacy including language, mathematics, and health; understanding scientific practices and recognizing science as a social process; and judging scientific expertise, dispositions, and ways of thinking and processing information. One important recommendation of the committee was that science literacy needs to be understood from a structural societal framework in which schools and other institutions might be enhancing or limiting individuals' access to science literacy, and that societies and communities have to take responsibility for science literacy beyond the individual: for example, by identifying unhealthy water filtration and its impact on individuals and the larger community's well-being.

Science learning cannot be remanded strictly to the schools; parents and communities also have an important role to play. We need to stress, in our homes, the importance of becoming more active in understanding the why and how of the world around us. Learning science means working with others to bring a usable knowledge to situations, using scientific practices to answer questions drawn from multiple disciplines.

One interesting example of how communities can address this responsibility is the Collaborative Science Education Centre at the University of Helsinki.[17] The center, founded in 2003, offers science programs from early childhood education through higher education. Families can bring their children to the center to participate in innovative educational experiences that are interdisciplinary, combining humanities (including art), mathematics, and natural sciences. The center works with private industries and with the city of Helsinki and nearby municipalities to highlight learning science environments and different approaches to learning, provide teaching materials, and offer courses for pre-kindergarteners through doctoral-level candidates. More than 500,000 children, youth, family members, teachers, and those interested in becoming teachers have participated in the creation and promotion of new approaches to research-based science education, both in-person

and virtually. The center also constructs science laboratories for schools and spon-
sors science clubs, science adventures, and camps as inspirational experiences for
discovering science knowledge and fundamental literacy. New courses are developed for teacher education that can be accessed by future teachers and practicing
teachers, all of whom learn and study together. The goal of the center is to contribute to the university's "social role" of promoting science literacy. Many of the National Academies' recommendations regarding science literacy are thriving in this
center—not just on paper, but also experientially, as our U.S. team clearly saw when
it visited on multiple occasions.

The best way to acquire knowledge about science is within an environment
that supports social and emotional learning. Life today occurs within social systems, whether in the home or the workplace. We now know a lot about what creates productive positive learning environments, and it is clear that the difficult
problems of the future will require collaboration, cooperation, listening to other
points of view, and figuring out explanations and solutions together. These positive learning environments support all of the other components of science learning: purposeful questions that spark and sustain interest; new skill acquisition and
knowledge development through the use of scientific practices; and levels of challenge and engagement that enhance feelings of success, confidence, and ownership of ideas and their solutions.

> Recommendation 5: To make science reform a reality for everyone, its
> implementation should involve students, their teachers, parents, and our
> entire shared, global community.

The world is going to be facing many difficult challenges in the years to come.
Because the challenges of today will not necessarily be the challenges of tomorrow, we need students who will be imaginative problem-solvers, envisioning and
creating innovative solutions. We cannot expect this to happen without seriously
making changes in how we teach science, especially in high school. We need students to leave school ready to understand scientific issues and take the necessary
actions—either as scientists or citizens—to promote the importance, value, and

necessity of scientific learning for finding solutions to new and old problems. PBL is one instructional method with a long history that fits closely with international recommendations for how to change the teaching and learning of science in schools. Our research shows its promise and we are committed to further testing its effects.

Appendix A

THE CRAFTING ENGAGEMENT IN SCIENCE ENVIRONMENTS STUDY

Crafting Engagement in Science Environments (CESE), funded by the National Science Foundation and the Academy of Finland, is a collaboration between researchers at Michigan State University and the University of Helsinki to increase student engagement and achievement in high-school physics and chemistry classes. Declining interest, achievement, and pursuit of careers in science, technology, engineering, and math (STEM) have become a global concern of educators, policymakers, and the general public, prompting several initiatives to improve science learning and instruction. Two influential documents have called for major reforms in science learning and instruction across the U.S. education system, the *Framework for K–12 Science Education* and *Next Generation Science Standards* (NGSS). These documents emphasize the importance of developing a deep understanding of scientific ideas and scientific practices over rote memorization of discrete science facts.

This new vision of scientific proficiency encourages students to make sense of phenomena and find solutions to problems using three-dimensional learning comprised of (1) core ideas from the scientific disciplines, (2) scientific and engineering practices, and (3) crosscutting concepts that span multiple areas of study. Crosscutting concepts are ideas that have application across different science domains. Comparable efforts in Finland resulted in guidelines similar in focus and content to those in the United States.

The aim of CESE is to help teachers and schools meet NGSS and achieve three-dimensional learning through the use of project-based learning (PBL). Our international team of researchers is working with teachers in the United States and Finland to create PBL units and test their impact on student science achievement and social and emotional learning. The U.S. team field-tested three physics and three chemistry units in Michigan high schools (complete with unit assessments and instructional materials). Participating teachers received ongoing professional learning and technical support to implement and refine the intervention. Schools in Finland are also using PBL materials from the United States, and are testing their impact among a dozen secondary school teachers who have also participated in professional development activities in the United States. Through these efforts and work in several other countries, CESE is engaged in developing a professional international science learning community.

To measure student engagement and social and emotional states, our project uses the experience sampling method (ESM), administered on smartphones equipped with an open source application specifically designed to capture what students are doing at specific moments and how they feel about it. Students are signaled several times during the day and prompted to answer a short survey about where they are, what they are doing, and who they are with, along with a number of questions related to their feelings at the moment when signaled. Science teachers involved in the project answer their own ESM surveys at the same time as their students, and both teachers and students complete surveys about their attitudes, backgrounds, and career goals. Classroom observations, video data, and student artifacts are also collected to measure the fidelity of implementation, and pre- and post-tests measure student science learning.

A special focus of this project is to bring high-quality science instruction to schools serving predominately low-income and minority students. Consequently, future evaluation efforts will consider both the average effect of the intervention and its differential effect on student subgroups. CESE has worked with over fifty teachers in twenty schools and served 1,400 students, more than a third of whom are low-income and minority. Several major papers and presentations have been made on the project, and the PBL units are currently being formatted for distribution by OECD as part of their new initiative on developing instruction that enhances creativity. All of our data are being archived at the Inter-university Consortium for Political and Social Research (ICPSR) and will be available for further research and replication. Data from this project will be available at doi.org/10.3886/E100380V1.

Appendix B

Unit Driving Question

How can I design a vehicle to be safer for a passenger during a collision?

Targeted NGSS Performance Expectations

Students who demonstrate understanding can:

HS-PS2–1: Analyze data to support the claim that Newton's second law of motion describes the mathematical relationship among the net force on a macroscopic object, its mass, and its acceleration.

HS-PS2–3: Apply scientific and engineering ideas to design, evaluate, and refine a device that minimizes the force on a macroscopic object during a collision.

Question(s)	Phenomena	Scientific and Engineering Practice(s)	What We Figured Out: Disciplinary Core Ideas (DCI), Crosscutting Concepts (CCC)	Lesson Level Three-Dimensional Learning Performances
1.1 What happens during a vehicle collision?	A series of two-body collisions	Asking questions and defining a problem	Students figure out the differences in the relationship between velocity and force in different collisions. Cause and effect	Students will generate questions about, and create initial models of, the relationship between velocity and force of a two-body collision.

Lesson Description: The teacher will present the driving question and explain the challenge to design a vehicle that is as safe as possible during a collision. Students will generate questions to determine what additional information they need to answer the driving question. Students will view a series of videos that show collisions between two bodies, and begin to identify important concepts that are being shown. Lastly, students will use toy cars to generate initial models of vehicle collisions, taking size and speed of the vehicles into account.

| 1.2 What variables affect force during a vehicle collision? | A series of two-body collisions | Analyzing and interpreting data

Planning and carrying out investigations | Students figure out the relationship between force, mass, and velocity of an object.

Cause and effect | Students will plan and carry out an investigation to determine the relationship between the mass and speed over time (acceleration) of objects and the force of collisions. |

Lesson Description: Students will brainstorm variables that can influence a collision. Students will collaboratively design an investigation using carts, ramps, various masses (bolts to place in the carts), and a wooden block to determine the relationships between the variables (force, mass, and velocity over time) during a collision.

Question(s)	Phenomena	Scientific and Engineering Practice(s)	What We Figured Out: Disciplinary Core Ideas (DCI), Crosscutting Concepts (CCC)	Lesson Level Three-Dimensional Learning Performances
1.3 How can we use a model to explain the relationship between the different components of a collision?	Collision of car and block	Analyzing data Using mathematics and computational thinking	Students figure out that when the mass or acceleration of an object increases, its force will increase (Newton's second law). Systems and system models	Students will use mathematical and computational thinking to analyze data by creating graphical representations of the relationships between force, mass, and acceleration. Students will use evidence to develop initial models explaining the relationships between force, mass, and acceleration.

Lesson Description: Students will use data from previous lesson's investigation to create line graphs measuring the relationship between variables. The class will discuss the accuracy/ precision of their data and ways to improve the validity of their results (through controlling more variables). Students will use their data to draw a model that answers the question: How do mass and acceleration affect the force of a collision?

Question(s)	Phenomena	Scientific and Engineering Practice(s)	What We Figured Out: Disciplinary Core Ideas (DCI), Crosscutting Concepts (CCC)	Lesson Level Three-Dimensional Learning Performances
1.4 How can I develop a model to explain a collision?	Collision of car and block.	Developing models	Students figure out that when the mass or acceleration of an object changes, the force of the collision will change. Systems and system models	Students will develop an interactive model to explain the relationship between the force of vehicle collisions and the mass or the speed/acceleration of the objects in the collision.

Question(s)	Phenomena	Scientific and Engineering Practice(s)	What We Figured Out: Disciplinary Core Ideas (DCI), Crosscutting Concepts (CCC)	Lesson Level Three-Dimensional Learning Performances

Lesson Description: Students will create a model to explain the relationship between the force of vehicle collisions and the mass or the speed/acceleration of the objects in the collisions using an interactive electronic modeling program called SageModeler. The students explore the simulation in order to gain a better working understanding of the program. Students will discuss the benefits of having different models to represent data.

1.5 How can you minimize force during a collision?	Tossing water balloons	Constructing explanations and designing solutions	Students figure out that force is minimized at a given moment when it is applied on an object over a longer period of time. Cause and effect	Students will investigate the relationship between the force and duration of a collision by tossing water balloons.

Lesson Description: Students will investigate, by tossing water balloons, how changing the amount of time a force is applied during a collision changes the force of the collision. Students will participate in a water-balloon-toss competition, then watch a slow-motion video of it to create a model to explain what happens during the collision.

1.6 How can I develop a model to explain the balloon investigation?	Impulse, relationship between variables	Constructing explanations and designing solutions Using mathematics and computational thinking	Students figure out that changes in the stopping speed or deceleration of objects during a collision change the force. Systems and system models	Students will revise a model to explain the relationship between the force of the collision and the stopping speed or deceleration of the objects in the collision using SageModeler.

Lesson Description: Students will revise their earlier models to explain the relationship between the force of collisions and the stopping speed or deceleration of the object in the collisions using SageModeler. Students will share their models and provide feedback to other groups.

Question(s)	Phenomena	Scientific and Engineering Practice(s)	What We Figured Out: Disciplinary Core Ideas (DCI), Crosscutting Concepts (CCC)	Lesson Level Three-Dimensional Learning Performances
1.7 How does changing the design of a cart affect its safety?	Collision of cart with block	Planning and carrying out an investigation	Students figure out that different materials can change the stopping speed or deceleration of the objects during a collision. Cause and effect	Students will plan and carry out an investigation to determine the effect that placing different materials on a cart has on the force of the collision.
Lesson Description: Students will connect the balloon-toss lesson to this initial force investigation by investigating and observing the effects that different materials have on the force of the collision when placed on the front of a cart.				
1.8 How can I design a vehicle that is as safe as possible?	Vehicle that has an egg "passenger"	Constructing explanations and designing solutions	Students figure out criteria and constraints involved in designing a solution to minimize the force during a collision. Structure and function	Students will design safety apparatuses to minimize the damage on a vehicle caused by the force of a collision.
Lesson Description: Students will brainstorm design constraints and materials to improve a car so that it will minimize the force on an egg during the impact of a collision. Students will develop solutions that change the makeup/design of the vehicle, such as adding clay or Play-Doh, or adding some sort of parachute to the back. Students will create a design sketch, then begin constructing their safety apparatus.				

Question(s)	Phenomena	Scientific and Engineering Practice(s)	What We Figured Out: Disciplinary Core Ideas (DCI), Crosscutting Concepts (CCC)	Lesson Level Three-Dimensional Learning Performances
1.9 How can I revise my vehicle design to make it as safe as possible?	Vehicle collision with egg as "passenger"	Constructing explanations and designing solutions	Students figure out that different combinations of solutions can change the stopping speed or deceleration of the objects during a collision. Patterns	Students will use evidence from data to revise designs for safety apparatuses to minimize the damage on a vehicle during a collision.

Lesson Description: Students will complete their cart designs/safety apparatus and test their revised cars during a collision. Students will make additional revisions when necessary, and add to their sketches/models to incorporate those changes.

Question(s)	Phenomena	Scientific and Engineering Practice(s)	What We Figured Out: DCI, CCC	Lesson Level Three-Dimensional Learning Performances
1.10 How are forces and motion related to vehicle safety?	Vehicle collisions	Explanations Communicating information	Students figure out the relationship between force, motion, and speed over time and vehicle safety. Cause and effect	Students will use models and data from investigations as evidence to communicate and explain the relationship between force, motion, and speed over time and vehicle safety.

Lesson Description: Students develop a presentation to discuss the relationship between vehicle safety and Newton's laws by focusing on one sub-question and describing how the class addressed the question and what conclusions we came to. Students will share artifacts from that lesson, create a personal reflection related to the lesson, and present their presentations, findings, and designs (using their models, design sketches, and cars) in a "carousel" format.

Appendix C

STUDENT AND TEACHER ESM QUESTIONNAIRES

Student ESM Questionnaire

I. GENERAL QUESTIONS

1. Where were you when you were signaled?
    a. Science Class
        i. Physics
        ii. Chemistry
        iii. Biology
        iv. Other
    b. Mathematics Class
    c. English Language Arts/Mother Language (in Finland)
    d. Social Studies Class
    e. Foreign Language Class
    f. Other class not listed
    g. In school but not in class
    h. Taking a break (Finland only)
    i. Out of school

2. What were you doing when signaled? (if Q1 = answers a–g)
    a. Listening
    b. Discussing
    c. Writing
    d. Calculating
    e. Taking a quiz/test
    f. Working on a computer
    g. Working in a group
    h. Laboratory work
    i. Presenting
    j. Other

3. Which best describes what you were doing in science when signaled? Check all that apply (if Q1 = a)
    a. Asking questions
    b. Defining problems
    c. Developing models
    d. Planning an investigation
    e. Conducting an investigation
    f. Analyzing data
    g. Interpreting data
    h. Solving problems
    i. Constructing an explanation
    j. Designing a solution
    k. Using evidence to make an argument
    l. Communicating information
    m. Other

4. What were you learning about in science when signaled? (if Q1 = a)
    [Open-ended]

5. Who were you with?
    a. Teacher
    b. Classmates
    c. Teacher and classmates
    d. Friends
    e. Other students
    f. Relatives
    g. Alone
    h. Other

6. Were you doing the main activity because you . . . ?
    [wanted to, had to, had nothing else to do]

7. Was what you were doing . . . ?
    [more like school work, more like play, both, neither]

II. HOW DID YOU FEEL ABOUT THE MAIN ACTIVITY?
(4-POINT SCALE: NOT AT ALL—VERY MUCH)

8. Were you interested in what you were doing?
9. Did you feel skilled at what you were doing?
10. Did you feel challenged by what you were doing?

11. Did you feel like giving up?
12. How much were you concentrating?
13. Did you enjoy what you were doing?
14. Did you feel like you were in control of what you were doing?
15. Were you succeeding?

III. HOW DID YOU FEEL ABOUT THE MAIN ACTIVITY?
(4-POINT SCALE: NOT AT ALL—VERY MUCH)

16. Was this important for you?
17. How important was this in relation to your future goals/plans?
18. Were you living up to the expectations of others?
19. Were you living up to your expectations?
20. I was so absorbed in what I was doing that the time flew.
21. How determined were you to accomplish the task?
22. When working on this activity . . . I used my imagination.
23. When working on this activity . . . I solved problems that had more than one possible solution.
24. When working on this activity . . . I explored different points of view on the problem or topic.
25. When working on this activity . . . I had to make connections with other school subjects.

IV. HOW DID YOU FEEL ABOUT THE MAIN ACTIVITY?
(4-POINT SCALE: NOT AT ALL—VERY MUCH)

26. Were you feeling . . . Happy
27. Were you feeling . . . Excited
28. Were you feeling . . . Anxious
29. Were you feeling . . . Competitive
30. Were you feeling . . . Lonely
31. Were you feeling . . . Stressed
32. Were you feeling . . . Proud
33. Were you feeling . . . Cooperative
34. Were you feeling . . . Bored
35. Were you feeling . . . Self-confident
36. Were you feeling . . . Confused
37. Were you feeling . . . Active

Teacher ESM Questionnaire

1. Which science practices were you emphasizing when signaled? Check all that apply.
   a. Asking questions
   b. Defining problems
   c. Developing models
   d. Planning an investigation
   e. Conducting an investigation
   f. Analyzing data
   g. Interpreting data
   h. Solving problems
   i. Constructing an explanation
   j. Designing a solution
   k. Using evidence to make an argument
   l. Communicating information
   m. Other

2. What were students doing in science when signaled?
   a. Listening
   b. Discussing
   c. Writing
   d. Calculating
   e. Taking a quiz/test
   f. Working on a computer
   g. Working in a group
   h. Laboratory work
   i. Presenting
   j. Other

3. How did you feel about the main activity? (4-point scale: Not at all—Very much)
   a. Were you feeling . . . Happy
   b. Were you feeling . . . Excited
   c. Were you feeling . . . Anxious
   d. Were you feeling . . . Competitive
   e. Were you feeling . . . Lonely
   f. Were you feeling . . . Stressed
   g. Were you feeling . . . Proud
   h. Were you feeling . . . Cooperative
   i. Were you feeling . . . Bored
   j. Were you feeling . . . Self-confident

k.   Were you feeling . . . Confused
l.   Were you feeling . . . Active

4.   Did you feel challenged by what you were doing?

5.   Were you interested in what you were doing?

6.   Did you feel skilled at what you were doing?

Appendix D

SINGLE-CASE DESIGN

Many ESM studies in schools have focused on describing students' experiences in situ, allowing investigators to compare how students are experiencing different situations or activities.[1] But if we want to move beyond just describing experience to assessing how changing the situational experiences of students affects them, we need to design experiments that will allow us to make comparisons between the instructional intervention and other situations. Although a traditional randomized trial with separate treatment and control groups would be ideal for making such comparisons, detecting the effects of an intervention at the classroom level requires a large, cluster-randomized trial.[2] ESM studies present considerable costs and logistical challenges that make executing such a large trial impractical for our initial work on this project. For example, a research team would need hundreds of smartphones and a large team of people to manage them in order to collect ESM data from both a treatment and control group of the size needed to detect effects.

Our study attempts to overcome some of these hurdles by combining the ESM with a single-case design to test the effects of the PBL treatment for high-school chemistry and physics classes on a variety of student outcomes.[3] In our design, teachers alternated between their regular instruction (baseline) and project-based units (treatment). Known as a reversal or ABAB design, this method establishes each classroom as its own experimental control through repeated replications of an effect.[4] In other words, during the baseline (A) phases, when teachers were engaged in "business-as-usual" instruction, we would expect to see lower levels of engagement, relevance, or imaginative thinking than during the treatment (B) phases. Demonstration of the expected changes in these levels between baseline and treatment phases, and replication of this pattern across classrooms, allows us to draw inferences about the effect of the intervention.

During each of the phases of the single-case design, students received three ESM signals a day, for three days, during their science class. Students also received five signals, delivered on a random schedule, throughout the rest of the day while they were not in their science class. Each time students were signaled, they answered survey questions about what they were doing when signaled, where they were, who they were with, and a number of questions asking them to rate their affect and experience. Within each science classroom, all students were signaled to answer the ESM survey at the same time in order to give us more information about what was happening at each moment and to minimize class disruptions.

One common method for the analysis of single-case designs is to graph the data for each case and to visually examine the changes between baseline and treatment phases.[5] Visual analysis is useful in studies when there are only a few cases, but statistical techniques have been developed for the synthesis of results when there are many cases within a study, as is the case in ours.[6] Because repeated measures are nested within students who are nested within classrooms, statistical techniques such as structural equation models (SEM) or hierarchical linear models (HLMs) are necessary for accurately estimating effects.[7]

For example, to estimate the effect of the PBL intervention on students' imaginative thinking, as in the results reported in Chapter 4, we used an HLM model with individual ESM responses (level 1) nested within students (level 2). We can also nest the students within their classrooms (level 3).

Level 1—ESM Responses

$$Y_{tij} = \beta_{0ij} + \beta_{1ij}T_{tij} + \beta_{2ij}X_{tij} + \varepsilon_{tij}$$

Level 2—Students

$$\beta_{0ij} = \gamma_{00j} + \gamma_{01j}Z_{ij} + \upsilon_{0ij}$$
$$\beta_{1ij} = \gamma_{10j}$$
$$\beta_{2ij} = \gamma_{20j}$$

Level 3—Classrooms

$$\gamma_{00j} = \delta_{000} + \delta_{001}W_j + \eta_{00j}$$
$$\gamma_{10j} = \delta_{100}$$
$$\gamma_{20j} = \delta_{200}$$

Y_{tij} is the momentary response for student i in classroom j at time t. T_{tij} is a binary indicator of whether the moment occurred during the baseline instruction or during the PBL treatment. The coefficient β_{1ij} contains the estimate of the overall treatment effect of PBL compared to business-as-usual instruction. X_{tij} is a series of moment-level covariates that might explain variation in students' experiences such as activity or scientific practice. At level 2, the student level, Z_{ij} is a vector of student characteristics that might explain variation in the outcome, such as gender, race or ethnicity, or baseline interest in science. Similarly, W_j represents a series of classroom-level characteristics such as teacher experience or the particular science course (physics or chemistry). For testing the single-case design, T_{tij} would be replaced by several indicators that compare each phase to the previous one, thus allowing us to see social and emotional variables rise and fall between each phase.[8]

The single-case design was adequate for the first stage of our study, but our goal is to thoroughly test PBL in high-school chemistry and physics classrooms in the United States and Finland. Our results thus far have been promising, but we recognize that if PBL is to be adopted widely, our test of its effects must meet more rigorous standards of evidence. So, over the next two years of our study we are scaling up to conduct a large, cluster-randomized trial which includes dozens of schools and thousands of students. Through this larger study we will be able to assess not only the overall impact of PBL, but also the differential impact on subgroups such as gender and race or ethnicity.

Notes

Why Learning Science Matters

1. The reference to the driving question will be explained later in this chapter. It is a key component of project-based learning (PBL), the intervention we are testing in our work. Additionally, it is important to underscore that this is an international study: in the chapters describing the Finnish experience, we follow the Finnish custom for referencing teachers. In Finland, students and their parents refer to teachers by their first names or nicknames, or simply as "teacher." There is no gender distinction in the Finnish language that specifically refers to "he" or "she."

2. See, e.g., European Commission, *Horizon 2020: Work Programme 2016–2017, Science with and for Society*; and European Commission, *Horizon 2020: Work Programme 2016–2017, European Commission Decision C(2017)2468*; and in the United States, NRC, *Framework for K–12 Science Education*; see also NGSS Lead States, *Next Generation Science Standards*.

3. Gago et al., *Europe Needs More Scientists*; OECD Global Science Forum, *Evolution of Student Interest in Science and Technology Studies*; Rocard et al., *Science Education Now*; and Osborne and Dillon, *Science Education in Europe*.

4. The 1950s have often been called the Golden Age of Science and the Sputnik era, which lasted through 1976. The initial federal allocation in 1957 to boost science education was 887 million U.S. dollars. During the 1980s a cycle of reports encouraged research, curriculum development, and teacher training in science and mathematics, but federal funding allocations were "topsy-turvy": new programs received support, followed years later by cuts, revisions, or terminations. From the 1990s to 2011, increased funding for science, technology, engineering, and mathematics (STEM) research and career training occurred across multiple federal entities. See Atkin and Black, *Inside Science Education Reform*. For an updated version, see Atkin and Black, "History of Science Curriculum Reform." Also see European Commission, *Horizon 2020: Work Programme 2016–2017, Science with and for Society*; and European Commission, *Horizon 2020: Work Programme 2016–2017, European Commission Decision C(2017)2468*; and in the United States, NRC, *Framework for K–12 Science Education*; and NGSS Lead States, *Next Generation Science Standards*. Europe has been moving in a similar direction—see Rocard et al., *Science Education Now*, 2006.

5. *Watch Mr. Wizard* started in 1951 and its popularity rapidly grew, with an estimated 800,000 viewers per episode. We mention *Watch Mr. Wizard* because the program featured experiments that could be done with boys and girls both at home and in the classroom. It was cancelled in 1965 and briefly revived in the 1970s. Mr. Wizard—Don Herbert—continued to develop experiments for schools and worked on a cable program for Nickelodeon. Reruns of the show ended in 2000. The question is, where is the Mr. Wizard of today? See *Watch Mr. Wizard*. From 1960 to 1966, *My Weekly Reader—The Children's Newspaper* published a series of imaginary space-like inventions, which soon materialized within a matter of years. These early classroom newspapers were designed to spark the imagination—a feature of science education that we have lost and need to regain.

6. National Academies of Sciences, Engineering, and Medicine, *Rising above the Gathering Storm*.

7. For information on Finnish students' declining interest in science-related careers, see Lavonen and Laaksonen, "Context of Teaching and Learning School Science in Finland." Additional information is shown in the PISA results from 2006–2015. The PISA framework emphasizes attitudes toward science including interest, enjoyment, and perceived value of science. Between 2006 and 2015, Finland experienced decreases in the proportion of students who say they enjoy acquiring new knowledge in science. At the same time, the United States has seen a slight increase on that same question but American students' academic scores have not increased appreciably. See OECD, *PISA 2006*; and OECD, *PISA 2015 Results*, vol. 1: *Excellence and Equity in Education*.

8. As of November 2017, nineteen states have adopted the NGSS. See, e.g., the National Science Teachers Association, ngss.nsta.org.

9. See NRC, *Framework for K–12 Science Education*, 2012.

10. The NRC defines scientific and engineering practices as the major practices that scientists employ and that engineers use to design and build systems; crosscutting concepts refers to links across scientific domains, and disciplinary core concepts refers to key science content knowledge (ibid., 30–31). See Chapter 1 for a fuller description of the components of three-dimensional learning and how we use it with PBL.

11. See NGSS Lead States, *Next Generation Science Standards*, 2013.

12. See FMEC, *Tulevaisuuden lukio;* also FNBE, *National Core Curriculum for Basic Education*.

13. The Finnish curriculum is currently being revised (outside of its usual sequence). It is important to underscore that Finland does not use "standards" but rather describes "aims and objectives" (not learning outcomes). Professors Katariina Salmela-Aro and Jari Lavonen have been appointed by the Ministry of Education to serve on a reform group that will be advising the government on a new law concerning secondary schools, which will be released sometime in 2019.

14. See, e.g., European Commission, *Horizon 2020: Work Programme 2016–2017, Science with and for Society;* and European Commission, *Horizon 2020: Work Programme 2016–2017, European Commission Decision C(2017)2468*.

15. See Lavonen, "Educating Professional Teachers through the Master's Level Teacher Education Programme in Finland"; Juuti and Lavonen, "How Teaching Practices Are Connected"; Lavonen, "Building Blocks"; and Lavonen and Juuti, "Science at Finnish Compulsory School." Teacher professionalism in Finland not only refers to the competence of individual teachers but also to their status. Professionalism depends on factors operating at the school and cultural level and within the broader educational policy environment. Important school-level factors include the nature of school leadership, norms of collaboration and reach and density of networks, and stakeholder partnerships. Cultural and education policy factors include the state-level context, including whether the country is following an accountability educational policy or whether it trusts teachers' instructional quality without frequent systematic inspection and testing. See Sahlberg, *Finnish Lessons;* Darling-Hammond and Lieberman, *Teacher Education around the World;* Korhonen and Lavonen, "Crossing School–Family Boundaries"; and Niemi, Toom, and Kallioniemi, *Miracle of Education*.

16. See Provasnik et al., *Highlights*.

17. See OECD, *PISA 2006;* and OECD, *PISA 2012 Results in Focus.*

18. Cheung et al., "Effective Secondary Science Programs."

19. See Condliffe et al., "Project-Based Learning."

20. See Krajcik and Shin, "Project-Based Learning"; Krajcik and Czerniak, *Teaching Science;* Krajcik et al., "Planning Instruction"; and Krajcik and Merritt, "Engaging Students in Scientific Practices." Another important advocate for project-based learning is the Buck Institute for Education (BIE), which has played a major role in PBL adoption over the past two decades. Three recent books advocating a PBL approach are: Markham, Larmer, and Rabitz, *Project-Based Learning Handbook;* Larmer, Mergendoller, and Boss, *Setting the Standard for Project Based Learning;* and Boss with Larmer, *Project Based Teaching.*

21. For Jari Lavonen, see Juuti, Lavonen, and Meisalo, "Pragmatic Design-Based Research"; Lavonen, "Building Blocks," 2013; Lavonen et al., "A Professional Development Project"; Lavonen et al., "Attractiveness of Science Education."

22. See Krajcik and Shin, "Project-Based Learning."

23. Core ideas are referred to as performance expectations in the NGSS because they define the phenomena that students should learn and the scientific practices students should employ to understand not just a phenomenon's meaning, but also its relevance and connections to other concepts.

24. Klager, Schneider, and Salmela-Aro, "Enhancing Imagination and Problem-Solving in Science."

25. Bell, "Using Digital Modeling Tools and Curriculum Materials."

26. For Katariina Salmela-Aro's most cited pieces on this topic, see Salmela-Aro et al., "Dark Side of Internet Use"; Symonds, Schoon, and Salmela-Aro, "Developmental Trajectories"; Little, Salmela-Aro, and Phillips, *Personal Project Pursuit;* and Salmela-Aro, "Personal Goals and Well-Being."

27. For Barbara Schneider, see Csikszentmihalyi and Schneider, *Becoming Adult;* Schneider and Stevenson, *Ambitious Generation;* Schneider et al., "Impact of Being Labeled"; and Schneider et al., "Transitioning into Adulthood."

28. For a description of situational interest, see Schneider et al., "Investigating Optimal Learning Moments"; and Hidi and Renninger, "Four Phase Model." For more general definitions of engagement, see Fredricks, Blumenfeld, and Paris, "School Engagement"; and Fredricks and McColskey, "Measurement of Student Engagement." An alternative approach to engagement is one that focuses on the relationships between teacher, students, and class content; see, for example, Corso et al., "Where Student, Teacher, and Content Meet." This focus can also be found in Daniel Quin's newer *Review of Educational Research* article; see Quin, "Longitudinal and Contextual Associations." While these two pieces identify some of the same interactive relational aspects of engagement that we do, they are not situational, nor do they include multiple motivational, social, and emotional factors that we consider essential for understanding engagement.

29. Csikszentmihalyi, *Flow.*

30. When fully engaged in an OLM, a student feels successful (along with other positive emotions); when confusion or boredom creeps in, however, engagement decreases. In addition to the inherent uncertainty involved in undertaking these kinds of challenges, there must also be recognizable value in persisting if the student is to succeed and stay engaged. Further discussion of our model can be found in Chapter 4, where we fully explain our measures and their relationship with one another.

31. For the process we are using for our assessments, see Schneider et al., "Developing Three-Dimensional Assessment Tasks." In the future, we will be administering a summative independent assessment to measure if PBL had a positive effect on student science achievement.

32. For additional information on single-case design, see Horner and Odom, "Constructing Single-Case Research Designs"; Kennedy, *Single-Case Designs;* Kratochwill, *Single Subject Research;* and Kratochwill and Levin, "Introduction."

33. See Bryk, "Support a Science of Performance Improvement."

34. The number of students who participated in the work, along with the type of analytics we employed, allow us to make several reliable statistical statements about the impact of PBL on their social, emotional, and cognitive learning. Nonetheless, what we are finding cannot and should not be viewed as generalizable to the entire population of U.S. and Finnish high-school students taking chemistry and physics, or their teachers.

35. Vincent-Lancrin, *Teaching, Assessing, and Learning Creative and Critical Thinking Skills.*

36. Bransford, Brown, and Cocking, *How People Learn;* and National Academies of Sciences, Engineering, and Medicine, *How People Learn II.* To assess whether our teachers and students were open to learning about PBL or had a fixed mindset about learning new approaches, we used questions that were developed by Jennifer Schmidt (and adopted from Carol Dweck) specifically for middle and secondary teachers and students. See Schmidt, Rosenberg, and Beymer, "Person-in-Context Approach." For the classic study on the long-term effects of growth mindset, see Blackwell, Trzesniewski, and Dweck, "Implicit Theories of Intelligence." For a more approachable primer to the concept of growth mindset, see Dweck, *Mindset.*

Chapter 1. Creating Science Activities That Engage and Inspire

1. In Finland, young people do not drive until they receive their driver's license when they are adults. The PBL professional learning classroom exercise described here is to show the value of the "driving question," which relates to forces and motion. Were this taking place in Finland, we could substitute the collision of two objects such as bicycles or sleds.

2. Kilpatrick, "Project Teaching"; and Kilpatrick, "Project Method."

3. For further discussion on Dewey, see Dewey, *Experience and Education;* Dewey and Small, *My Pedagogic Creed.* Additional commentary on the differences between Kilpatrick and Dewey can be found in Larmer, Mergendoller, and Boss, *Setting the Standard;* for more on the failures of Kilpatrick and his influence on the progressive education movement, see Knoll, "I Had Made a Mistake."

4. Some critiques of PBL have reached into historical philosophy for a rationale of learning through projects. See, e.g., Condliffe et al., "Project-Based Learning." We take a more contemporary view, especially since many of Dewey's ideas—see Dewey, *Experience and Education*—resonate more closely with the PBL literature of the 1990s; see, e.g., Blumenfeld et al., "Motivating Project-Based Learning."

5. Knoll, "I Had Made a Mistake."

6. Savin Baden and Howell Major, *Foundations of Problem-Based Learning.* Savin Baden and Howell Major's work traces problem-based learning to classical Greek ideas—we have chosen to highlight its more contemporary origin given how it is typically reviewed in the European education literature and reports; see, for example, Sjøberg and Schreiner, *ROSE Project.*

7. Rocard et al., *Science Education Now.*

8. Osborne and Dillion, *Science Education in Europe.*

9. Lavonen, "Influence of an International Professional Development Project"; and FMEC, *Tulevaisuuden lukio.*

10. For connections between learning science theories and recommendations made in the

NGSS, see Edelson and Reiser, "Making Authentic Practices Accessible." See Design-Based Research Collective, "Design-Based Research." Design-based research has had both proponents and detractors: this article highlights the characteristics of good design-based research. PBL should be considered a design-based intervention that is embedded in theory, undergoes continuous cycles of redesign, functions in authentic settings, and uses methods that can be documented and are testable. Design-based research sets the stage for the systematic, randomized trials that we are in the process of conducting.

11. See Blumenfeld et al., "Motivating Project-Based Learning."

12. One of the major problems emphasized in Thomas's review (and by others) is that the "project" is the central teaching strategy and students learn the central concepts of the discipline via the project; Thomas, *Review of Research*. We argue that this problem arises because in many of the earlier cases using this method, the connections between performance expectations and learning goals were not well defined. PBL is not merely learning by doing; the learning must be embedded in a theory, with clear learning goals that wrap around a strategy—one driven by a question that creates learning activities involving scientific practices that can then be tested with an assessment process.

13. See Condliffe, "Project-Based Learning," 2017. Another synthesis review has been conducted by our team member and research associate, Christopher Klager—in his working paper, Klager identified only the science studies that actually reported that they were assessing PBL interventions in science. He discusses the difficulty of distinguishing between studies that just use projects in science instead of those using the criteria described earlier. Albeit with a small sample, he finds a positive effect with a medium effect size. Klager, "Project-Based Learning in Science."

14. See Kastberg et al., *Performance;* and Mullis et al., *TIMSS Advanced 2015 International Results.* For an interesting piece on the need for science literacy, see Miller, "Conceptualization and Measurement of Civic Scientific Literacy." Miller makes a strong case for acquiring tools to make sense of science, rather than learning details about soon-to-be-replaced "current" science terms and constructs, especially as students move through higher education. The issue regarding labor market trends can be found in the NSB's *Science & Engineering Indicators 2018.*

15. See NGSS Lead States, *Next Generation Science Standards.*

16. For more on this point in Finland, see Lavonen, "Influence of an International Professional Development Project." For the United States, see Krajcik and Shin, "Project-Based Learning."

17. It is important to note, for purposes of clarity, that inquiry-based science education (IBSE) has been suggested as a solution for engaging students in science learning—and, like PBL, has been criticized as being ill-defined and incomplete in its design and measures. From our Finnish colleagues, we have learned that the science standards movement in Finland and the United States are similar (this holds true for many other countries, as well, where it is typically referred to as IBSE). They have chosen to use PBL and scientific practices as a framework rather than IBSE. For more on this point, see Lavonen, "Influence of an International Professional Development Project"; and FMEC, *Tulevaisuuden lukio.*

18. A deeper discussion of the field test design can be found in Chapter 4.

19. See Lavonen, "Influence of an International Professional Development Project."

20. A number of researchers have contributed to the PBL literature. See, e.g., Kanter, "Doing the Project and Learning the Content"; Mergendoller, Maxwell, and Bellisimo, "Effectiveness of Problem-Based Instruction"; and Bell, "Project-Based Learning." One of the earliest researchers involved in describing the elements and benefits of PBL was Phyllis Blumenfeld: see Blumenfeld

et al., "Motivating Project-Based Learning." For an additional earlier publication, see Krajcik and Blumenfeld, "Project-Based Learning."

21. These factors are taken from Krajcik and Shin, "Project-Based Learning"; and Krajcik and Czerniak, *Teaching Science.* Also see McNeill and Krajcik, *Supporting Grade 5–8 Students.*

22. NGSS Lead States, *Next Generation Science Standards.*

23. See Jacob et al., "Are Expectations Alone Enough?"

24. See Hinojosa et al., *Exploring the Foundations of the Future STEM Workforce.*

25. The Finnish curricular version we used for aligning with the U.S. NGSS was published in 2014 and 2015. The Finnish national curriculum describes broad goals for learning and subject-specific core aims; however, municipal authorities have autonomy so curriculum is locally driven. Similar to PISA, the national curriculum uses scientific conceptual and procedural knowledge to explain phenomena, evaluate and design science inquiry, and interpret data and evidence scientifically. See Vitikka, Krokfors, and Hurmerinta, "Finnish National Core Curriculum"; OECD, *PISA 2015 Draft Science Framework;* and Lavonen, "National Science Education Standards and Assessment in Finland."

26. In Finland, teacher-awarded grades are used as recommendations but are not binding, and students have real choices when making decisions about their secondary school experience. Both pathways can lead to postsecondary school, but the majority of those attending vocational school tend to enter the labor market. See Lavonen, "Influence of an International Professional Development Project."

27. For responses by Finnish students, see Lavonen and Laaksonen, "Context of Teaching and Learning School Science in Finland."

28. See Juuti and Lavonen, "How Teaching Practices Are Connected."

29. Ibid.

30. See Funk and Hefferon, "As the Need for Highly Trained Scientists Grows." Also see Chow and Salmela-Aro, "Task Values across Subject Domains."

31. In Finland, the problem has primarily been centered around females choosing not to enter STEM fields or newly emerging industries. For the U.S. perspective, see NSB, *Science & Engineering Indicators 2018.*

32. See Juuti and Lavonen, "How Teaching Practices Are Connected"; and their most recent work on assessment, Lavonen and Juuti, "Evaluating Learning."

33. Examples of the storyline and daily lesson-level plans can be found in Chapter 2. Also see Bielik et al., "High School Teachers' Perspectives." Other U.S. team members who had active roles in the development of the chemistry units include Deborah Peek-Brown and Kellie Finnie; Peek-Brown was also a lead on the physics units.

34. The NGSS performance expectations numbering system can be found in *How to Read the Next Generation Science Standards,* available at https://www.nextgenscience.org/resources/how-read-next-generation-science-standards. The first digit indicates the grade level; in our case they all begin with "HS" because our units are for secondary school. Next comes the code to indicate the science discipline (in our case "PS" for physical sciences), then a number to indicate the disciplinary core idea followed by the order in which they appear in the framework.

35. For the latest results on our modeling work, see Klager, Chester, and Touitou, "Social and Emotional Experiences of Students."

36. See Lavonen and Juuti, "Evaluating Learning."

37. See Harris et al., *Constructing Assessment Tasks.*

38. Touitou et al., "Effects of Project-Based Learning."

39. The paradigm we use in the development and assessment of our professional learning

activities in Finland is described in Lavonen, "Building Blocks"; and in the United States, in Krajcik and Czerniak, *Teaching Science*.

40. Bielik et al., "High School Teachers' Perspectives."

41. See Lavonen, "International Professional Development Project."

42. For Finland, see Lavonen and Laaksonen, "Context of Teaching and Learning School Science in Finland." For the United States, see Krajcik, "Project-Based Science"; and Krajcik et al., "Planning Instruction."

Chapter 2. Project-Based Learning in U.S. Physics Classrooms

1. Mr. Cook represents several teachers who have been involved in the development and enactment of the physics unit. The portrayal of Mr. Cook was created by Deborah Peek-Brown, our lead curricular developer. Peek-Brown worked on both the physics and chemistry units and conducted observations of and video-recorded many of the classroom activities. The pictures of the students included in this chapter are stills from videos of one physics classroom during the forces and motion unit.

2. See Graham et al., *Dive In!*; and McNeill, Katsh-Singer, and Pelletier, "Assessing Science Practices." Another earlier practical guide for science teachers that emphasizes "inquiry-based science" can be found in Llewellyn, *Teaching High School Science*.

3. NRC, *Guide to Implementing the Next Generation Science Standards*. Also see Chopyak and Bybee, *Instructional Materials and Implementation of Next Generation Science Standards*.

4. The description of the typical physics unit on forces and motion was abstracted from the reflections of the teachers in our study on how physics is typically taught—including the reference to Newton's second law of motion (force equals mass times acceleration).

5. The traditional approach described earlier helps students solve textbook problems that are often used in examinations. We argue that PBL can help to solve problems because it identifies a phenomenon made "relevant" to real-life experiences, which students learn to explain through scientific and engineering practices that can be applied when solving similar and other types of common textbook problems. Although we have not yet tested the generalizability of this claim, in our work measuring student academic performance over time, we find that students' academic performance during PBL activities significantly raises their performance during more traditional science instruction, in both the United States and Finland (see Chapter 4).

6. In the classes detailed in Chapters 2 and 3, we had teacher and student permission to use video-captured pictures for our research and this book. The pictures shown in the lessons were extracted from videos that were taken over several weeks and represent some of the learning progressions that occurred during the PBL lessons. This process enabled documentation of teacher and student narratives happening in "real time." The pool of videos that had longitudinal components with clearly audible teacher and student voices that could be transcribed was limited. Mr. Cook's and Elias's teacher/student classroom examples most closely fit these criteria.

7. The driving question board is a visual organizing tool for PBL units. For more on the driving question board and the "activity summary board"—a complementary tool that can be used in combination with the driving question board to help students organize their roles and activities— see Touitou et al., "Activity Summary Board."

8. Here we have specified this activity as occurring at days four and five in the unit. But it may occur later, depending on how long the teacher takes to complete different activities. The basic idea here is to show the beginning, middle, and end of the unit.

9. SageModeler is an online tool that allows students to create and revise representations of their models throughout the unit. Students often begin a unit with some ideas about how to explain a phenomenon, but they often have misconceptions or understand only part of the underlying relationships. As the unit progresses and students gather more information, they update their models to better reflect how phenomena happen. See an example of a final model that a student created once they had investigated forces and motion in their physics class at https://concord.org/our-work/research-projects/building-models/. See also Bielik, Damelin, and Krajcik, "Why Do Fishermen Need Forests?"; Damelin et al., "Students Making System Models"; and Klager, Chester, and Touitou, "Social and Emotional Experiences of Students."

10. Our choice of the word "imaginatively" is deliberate, because it is one of the measures we use in our ESM measurement. We ask students about how imaginative they felt during specific PBL science experiences and during more conventional science lessons. See Klager, Schneider, and Salmela-Aro, "Enhancing Imagination and Problem-Solving in Science."

11. Linnansaari et al., "Finnish Students' Engagement in Science Lessons."

12. Upadyaya et al., "Associations."

Chapter 3. Project-Based Learning in Finnish Physics Classrooms

1. We use Mr. Falck's first name in this chapter, because this is the custom in the Finnish education system.

2. The Finnish framework curriculum for high-school basic mechanics emphasizes learning models that present motion with constant velocity and motion with constant acceleration. It also emphasizes that two forces result when two bodies interact: the first object acts on the second and the second on the first. In an interaction, the object with a larger mass will experience a smaller acceleration and the smaller object a larger acceleration. The curriculum introduces distance forces, weight, and magnetic force, and touch forces like friction, normal force, and air resistance.

3. If a state adopts or even adapts the NGSS, then U.S. students in these states would be required to develop understandings of these ideas as well.

4. See NGSS Lead States, *Next Generation Science Standards*. See also FMEC, *Tulevaisuuden lukio*; and FNBE, *National Core Curriculum for Basic Education*.

5. Dumont, Istance, and Benavides, *Nature of Learning*.

6. These findings are presented in OECD, *PISA 2015 Results*, vol. 1: *Excellence and Equity in Education*.

7. See Lavonen and Laaksonen, "Context of Teaching and Learning School Science in Finland."

8. FMEC, *Kiuru*.

9. FNBE, *National Core Curriculum for Basic Education*; FNBE, *National Core Curriculum for Upper Secondary Education*; and FMEC, *Tulevaisuuden lukio*.

10. Lavonen, "Educating Professional Teachers in Finland."

11. Aksela, Oikkonen, and Halonen, *Collaborative Science Education*.

12. Rushton et al., "Towards a High Quality High School Workforce."

13. See Schneider, Chen, and Klager, "Gender Equity and the Allocation of Parent Resources."

14. These results show the overall engagement for the students in both countries over the course of a year.

15. Touitou and Krajcik, "Developing and Measuring Optimal Learning Environments through Project-Based Learning."

16. Hietin, "2 in 5 High Schools Don't Offer Physics."

Chapter 4. How Learning Science Affects Emotions and Achievement

1. See National Academies of Sciences, Engineering, and Medicine, *How People Learn II.*

2. For information on surveys of science interest, see ongoing work by the Pew Research Center (http://www.pewinternet.org/2015/01/29), specifically Funk and Rainie, "Public and Scientists' Views on Science and Society."

3. NRC, *Framework for K–12 Science Education;* NGSS Lead States, *Next Generation Science Standards;* FNBE, *National Core Curriculum for Basic Education;* and FNBE, *National Core Curriculum for Upper Secondary Education.* We pay particular attention to how students' experiences differ by gender, because females have typically been less likely to take advanced-level science courses or to enter and persist in specific types of STEM careers. See Perez-Felkner et al., "Female and Male Adolescents' Subjective Orientations"; and Schneider et al., "Does the Gender Gap in STEM Majors Vary by Field and Institutional Selectivity?"

4. See Immordino-Yang, *Emotions, Learning, and the Brain.*

5. See Fredricks and McColskey, "Measurement of Student Engagement"; Linnenbrink-Garcia, Patall, and Pekrun, "Adaptive Motivation and Emotion in Education"; Tuominen-Soni and Salmela-Aro, "Schoolwork Engagement and Burnout"; and Inkinen et al., "Science Classroom Activities."

6. See Schneider et al., "Investigating Optimal Learning Moments." Also see Salmela-Aro et al., "Integrating the Light and Dark Sides of Student Engagement."

7. Csikszentmihalyi, *Flow.*

8. See Csikszentmihalyi and Schneider, *Becoming Adult;* Schmidt, Rosenberg, and Beymer, "Person-in-Context Approach"; Schneider and Stevenson, *Ambitious Generation;* and Shernoff et al., "Student Engagement."

9. Shumow, Schmidt, and Kacker, "Adolescents' Experience Doing Homework."

10. For a complete list of the questions asked on the ESM, see Appendix C.

11. Csikszentmihalyi, *Flow.*

12. See Duckworth, *Grit.*

13. See Dewey, *School and Society.*

14. See OECD, *PISA 2015 Results,* vol. 1: *Excellence and Equity in Education;* also see Lucas, Claxton, and Spencer, "Progression in Student Creativity in School." Their definition does not identify creativity as a trait but rather as a concept that is complex and multifaceted; happens in many dimensions of human life; is learnable; can be analyzed at an individual level; is influenced by context and social factors; and is identified as being desirable for various learning activities today.

15. Polman, *Designing Project-Based Science.*

16. The numbers referred to in the text use data collected between 2015 and 2018 across both the United States and Finland. Including data from the early pilot of our study in 2013 and 2014 as well, we have collected nearly fifty thousand responses from roughly 1,700 students. The studies reported in this section, however, use analytic samples from 2015–2016 and 2016–2017. Different studies have different sample sizes depending on when the studies were written and what data were available at the time they were submitted for publication. The reader can refer to the individual studies for more details on which years, countries, students, and responses were included in the analyses.

17. Schneider et al., "Investigating Optimal Learning Moments."

18. Additional results can be found in Spicer et al., "Conceptualization and Measurement of Student Engagement in Science."

19. Upadyaya et al., "Associations."

20. Schneider et al., "Investigating Optimal Learning Moments."

21. Linnansaari et al., "Finnish Students' Engagement in Science Lessons."

22. Moeller et al., "Does Anxiety in Math and Science Classrooms Impair Motivations?" For further definitions of burnout, see Salmela-Aro et al., "School Burnout Inventory."

23. See Salmela-Aro et al., "Integrating the Light and Dark Sides of Student Engagement." For a more developmental approach on many of these topics, see Salmela-Aro et al., "Dark Side of Internet Use."

24. Additional citations on burnout by Professor Salmela-Aro can be found in Chapter 5. One of her studies with Sanna Read combines the study of engagement and burnout profiles; see Salmela-Aro and Read, "Study Engagement and Burnout Profiles."

25. Salmela-Aro et al., "Does It Help to Have 'Sisu'?"

26. Duckworth, *Grit.*

27. Inkinen et al., "Science Classroom Activities and Student Situational Engagement."

28. See Shernoff, Knauth, and Makris, "The Quality of Classroom Experiences."

29. Horner and Odom, "Constructing Single-Case Research Designs"; Kennedy, *Single-Case Designs for Educational Research;* Kratochwill, *Single Subject Research;* Kratochwill and Levin, "Introduction."

30. Klager et al., "Creativity in a Project-Based Physics and Chemistry Intervention"; and Klager and Inkinen, "Socio-emotional Experiences of Students in Science."

31. Klager et al., "Creativity in a Project-Based Physics and Chemistry Intervention"; Klager and Schneider, "Enhancing Imagination and Problem-Solving Using Project-Based Learning."

32. It is important to note that our methodology is not robust enough at this stage to assume causation. In Year 2, however, we (1) sampled a different student population (same grade level and similar demographics); (2) sampled a different set of teachers (who also experienced professional development similar in content and experiences with those in Year 1); (3) used the same PBL units and scientific practices; (4) administered the same instruments and measures; and (5) conducted the same analytic procedures—with our findings from Year 2 being similar to those in Year 1 and of a stronger magnitude. It is important to recognize that our work to date includes (in the United States and Finland) 60 teachers, 24 schools, and 1,700 students. This is promising, but as Hedges cautions, our methodology does not warrant causal claims: see Hedges, "Challenges in Building Usable Knowledge in Education."

33. Klager and Schneider, "Strategies for Evaluating Curricular Interventions."

34. See NSF, *Preparing the Next Generation.*

35. See Luft and Hewson, "Research on Teacher Professional Development Programs in Science"; and Lavonen, "Building Blocks for High-Quality Science Education."

Chapter 5. Teachers Reflect on Project-Based Learning Environments

Epigraph. All of the quotations used in this chapter are from individual teachers in the United States and Finland. Names have been changed to protect anonymity. They have been edited slightly for readability.

1. References in this chapter are not intended to be a full analysis of teaching and learning in science. Much of what we feature with respect to teacher practices and student learning in science is shaped by the existing literature and the work of our principal investigator, Professor Lavonen. Here we have highlighted the ideas of Bransford that specifically address several of the measures we used to examine teacher and student engagement in learning science. *How People Learn: Brain, Mind, Experience, and School,* of which Bransford is the lead author, is used in the University of Helsinki teacher education program; see Bransford, Brown, and Cocking, *How Peo-*

ple Learn. Lavonen provides a complementary interpretation of Bransford's ideas, suggesting that active learning involves students' planning and evaluating their own learning. For a fuller discussion, see Osborne, Simon, and Collins, "Attitudes Towards Science."

2. These ideas are taken from Bransford, Brown, and Cocking, *How People Learn.*

3. For example, students report in the Programme for International Student Assessment (PISA) that this is the type of instruction they often receive. See OECD, *PISA 2015 Results,* vol. 3: *Students' Well-Being.*

4. In Finland, similar to the design principles of PBL, inquiry is defined as allowing for student freedom, reflection, interpretation, and evaluation through a variety of mediums including nature, web-based information, and conducting investigations. Observation is not just simply looking at things. The Finns recognize that observations are influenced by conceptions and beliefs, and that in science observations are used to generate explanations and theories about observed phenomena.

5. Finnish teachers have a different view of their relationship to national policies, which has been attributed to a variety of factors, including the cultural and historical value system whereby teachers feel they are active experts in the development of their high-quality education programs. See Sahlberg, *Finnish Lessons.*

6. The relationship between the new standard reforms in the United States and Finland and the response of the professional community are discussed in greater detail in the opening chapter.

7. Although not definitive, the ideas expressed by these teachers are nonetheless compelling and were widely shared among teachers in both countries. These interviews were conducted with twenty-five teachers, by several different interviewers. The framing of the excerpts from the interviews is based on a protocol given to teachers in both countries that was subsequently transcribed. We are interested in teachers' motivations to join the project so we can better understand potential selection bias and prior experience with PBL. Substantively, we probed about the teachers' knowledge and interpretation of three-dimensional learning, including disciplinary core ideas, crosscutting concepts, and scientific practices. Several specific design principles of PBL—such as the driving question, planning and carrying out investigations, collaborating to find solutions, and producing artifacts—were also explored. Additionally, we were interested in measuring student interactions in the lessons with regard to perceived academic ability (by the teacher), and in teachers' reflections on their PBL experiences. See Krajcik et al., "Planning Instruction"; and for a more specific in-depth explanation of PBL practices, see Krajcik and Czerniak, *Teaching Science.*

8. See Oikkonen et al., "Pre-Service Teacher Education"; and Niemi, "Educating Student Teachers."

9. For a description of the connection between personal and value-related interests in science, see Lavonen and Laaksonen, "Context of Teaching and Learning School Science in Finland."

10. In their newest book, Ryan and Deci point out that one of the key elements of a successful learning environment is the fostering of self-determination, especially when there is support (their word here is "compassion") for curiosity, creativity, and productivity. See Ryan and Deci, *Self-Determination Theory.*

11. There are a number of new handbooks on creativity. One of these is Shalley, Hitt, and Zhou, *Oxford Handbook.* We have highlighted this book because it focuses on the relationships between creativity and problem-solving, much of it within organizations. We think it is important to draw attention to the relationships among collaborative group problem-solving, activities, and artifact production, because this lays the groundwork for developing skill sets that are part of innovative entrepreneurial activity in multiple science fields, including technology and engineering. These ideas are also emphasized in the Finnish and U.S. science standards.

12. Klager, Chester, and Touitou, "Social and Emotional Experiences of Students."

13. Ibid.

14. Teacher professional development and the country teacher exchange are described in greater detail in Chapter 1.

15. To learn how the driving question and other dimensions of PBL relate to design principles and are constructed and evaluated in classroom practices, see Krajcik and Czerniak, *Teaching Science*, 2013.

16. Ibid.

17. These descriptions of scientific practices are abstracted from the NGSS. Collaboration is not identified in the NGSS as a scientific practice per se, but it is emphasized as a fundamental component of three-dimensional learning.

18. See Peek-Brown et al., "Using Artifacts."

19. NRC, *A Framework for K–12 Science Education*.

20. Klager and Schneider, "Strategies for Evaluating Curricular Interventions."

21. OECD, *TALIS 2013 Results*; Dweck, *Mindset*.

22. These numbers were calculated using standardized scores, meaning that we took into account differences in the individual mean score for each of the teachers and students.

23. These measures come directly from the ESM, "developing models and communicating information."

24. There have been concerns regarding the ESM methodology that have been addressed by several different economists, methodologists, psychologists, and sociologists. Taking into account missing data and other measurement problems found in surveys, the ESM continues to be a robust tool for measuring social and emotional learning. See Hektner, Schmidt, and Csikszentmihalyi, *Experience Sampling Method*.

25. NCES, *Schools and Staffing Survey*.

26. For a deep discussion on issues related to burnout both in other countries and in Finland, see work by our principal investigator Katariina Salmela-Aro, an expert in this area: Pietarinen et al., "Reducing Teacher Burnout"; Pyhältö, Pietarinen, and Salmela-Aro, "Teacher–Working-Environment Fit"; and Pietarinen et al., "Validity and Reliability of the Socio-Contextual Teacher Burnout Inventory."

27. See Shadish, Cook, and Campbell, *Experimental and Quasi-Experimental Designs*; and Schneider et al., *Estimating Causal Effects*.

28. The assessment process for a project as complex as this requires multiple procedures and accommodations into existing country-specific practices as well as new development activities. In Michigan, investigators will be partnering with testing personnel at the state level as well as with other states, developing summative test items that can accurately measure three-dimensional learning goals and activities. Working in collaboration with the Michigan team, Finland is also developing items that correspond to the units and customary practice in that country. Both of these item development processes are described in new papers by team members: see Bielik, Touitou, and Krajcik, "Crafting Assessments"; and Lavonen and Juuti, "Evaluating Learning."

29. Bryk and Schneider, *Trust in Schools*.

Chapter 6. Encouraging Three-Dimensional Learning

1. See Schneider et al., "Investigating Optimal Learning Moments." For further explanations on the value of social and emotional learning and its potential impact on school-based education instructional practices and policies, see Jones and Doolittle, "Social and Emotional Learning";

and Yeager, "Social and Emotional Learning Programs for Adolescents." Yeager casts some reservations on programs designed for this purpose and instead supports the idea that these behaviors are perhaps best nurtured in learning environments that create "climates and mindsets that help adolescents cope more successfully with the challenges they encounter." In such cases, "the evidence is not only encouraging" but actionable by schools and teachers (89).

2. Bransford, Brown, and Cocking, *How People Learn;* and National Academies of Sciences, Engineering, and Medicine, *How People Learn II.*

3. See Schneider et al., "Investigating Optimal Learning Moments."

4. See Salmela-Aro and Upadyaya, "School Burnout and Engagement."

5. For more on these features of PBL, see Chapters 1 and 2.

6. Bryk et al., *Learning to Improve.*

7. In securing the cooperation to enter the classrooms, we met with district and school leadership and science directors. One of the most affirming aspects of our work was their enthusiasm for PBL and their cooperation and expectation for involving more teachers and hopes for the development of additional units. They also were willing to share additional administrative records that we could use in the analyses of our data. This is also the case in Finland, where not only were the teachers and principals willing to collaborate, but the Ministry of Education was also excited with our work and its eventual scale-up.

8. We emphasize the importance of building a system of learning based on design principles. For further explanation of what constitutes design-based implementation research, see Fishman et al., "Design-Based Implementation Research."

9. See FMEC, *Development Programme for Teachers' Pre- and In-Service Education.* Also see Lavonen, "Educating Professional Teachers in Finland."

10. See TeachingWorks, "Deborah Lowenberg Ball—Director."

11. Several PBL units have been submitted to Achieve's Instructional Materials Review. This team of educators evaluates the quality of instructional products and is designed to identify high-quality materials that align with the NGSS (and Common Core State Standards [CCSS]). Several of the units designed by Krajcik and colleagues have been evaluated and found to best illustrate the cognitive demands of the NGSS. See www.achieve.org.

12. See Cobb et al., *Systems for Instructional Improvement;* although written based on middle-school mathematics, many of its recommendations reaffirm our approach to professional learning.

13. See ALLEA, *European Code of Conduct for Research Integrity,* which specifies standards of research integrity for projects that will be funded by Horizon 2020. (Horizon 2020 is the European Union Research and Innovation program promising breakthroughs and discoveries that take ideas from the lab to the market; see European Commission, *Horizon 2020: Work Programme 2016–2017, European Commission Decision.*)

14. See NGSS Lead States, *Next Generation Science Standards;* and Lee and Buxton, *Diversity and Equity in Science Education.*

15. Nearly a decade ago, Achieve published a report that examined ten sets of international science standards with the intent of informing a framework for the new U.S. science standards; see Achieve, *International Science Benchmarking Report.* This report highlighted variation among countries with respect to knowledge and skills and helped to set the stage for the NRC reports, the NGSS, and new standards that have emerged in other countries. The work of Achieve, the OECD, Finland, and other countries has indeed started a wave of science reform—the work for the next years ahead will undoubtedly be to develop curricular material, experiential activities, and assessment tools that can ensure that students are engaged in doing science. PBL offers a blue-

print for this next stage in science reform, and several countries—including Chile, China, and Israel—have approached our team regarding our secondary school units.

16. Snow and Dibner, *Science Literacy*.

17. Aksela, Oikkonen, and Halonen, *Collaborative Science Education*.

Appendix D

1. For a fuller description of potential uses of ESM studies, see Hektner, Schmidt, and Csikszentmihalyi, *Experience Sampling Method*.

2. Bloom, "Randomizing Groups."

3. Kennedy, *Single-Case Designs for Educational Research*; Kratochwill, *Single Subject Research*; and Kratochwill and Levin, "Introduction."

4. Horner and Odom, "Constructing Single-Case Research Designs."

5. Lane and Gast, "Visual Analysis."

6. Baek et al., "Use of Multilevel Analysis."

7. Muthén and Muthén, *Mplus User's Guide*; Raudenbush and Bryk, *Hierarchical Linear Models*.

8. Rindskopf and Ferron, "Using Multilevel Models."

Bibliography

Achieve. *International Science Benchmarking Report: Taking the Lead in Science Education; Forging Next-Generation Science Standards.* Washington, DC: Achieve, Inc., September 2010.

Aksela, Maija, Juha Oikkonen, and Julia Halonen, eds. *Collaborative Science Education at the University of Helsinki since 2003: New Solutions and Pedagogical Innovations for Teaching from Early Childhood Education to Universities.* Helsinki: University of Helsinki, 2018.

ALLEA (All European Academies). *European Code of Conduct for Research Integrity.* Berlin: ALLEA, 2017.

Atkin, J. Myron, and Paul Black. "History of Science Curriculum Reform in the United States and the United Kingdom." In *Handbook of Research on Science Education,* edited by Sandra K. Abell and Norman G. Lederman, 781–806. New Jersey: Lawrence Erlbaum, 2007.

———. *Inside Science Education Reform: A History of Curricular and Policy Change.* New York: Teachers College Press, 2003.

Baek, Eun Kyeng, Mariola Moeyaert, Merlande Petit-Bois, S. Natasha Beretvas, Wim Van den Noortgate, and John M. Ferron. "The Use of Multilevel Analysis for Integrating Single-Case Experimental Design Results within a Study and across Studies." *Neuropsychological Rehabilitation* 24, nos. 3–4 (2014): 590–606.

Bell, Stephanie. "Project-Based Learning for the 21st Century: Skills for the Future." *Clearing House* 83, no. 2 (2010): 39–43.

———. "Using Digital Modeling Tools and Curriculum Materials to Support Students' Modeling Practice." Paper presented at the annual meeting of the American Educational Research Association, San Antonio, TX, April/May 2017.

Bielik, Tom, Daniel Damelin, and Joseph Krajcik. "Why Do Fishermen Need Forests? Developing a Project-Based Learning Unit with an Engaging Driving Question." *Science Scope* 41, no. 6 (2018): 64–72.

Bielik, Tom, Kellie Finnie, Deborah Peek-Brown, Christopher Klager, Israel Touitou, Barbara Schneider, and Joseph Krajcik. "High School Teachers' Perspectives on Shifting towards Teaching NGSS-Aligned Project-Based Learning Curricular Units." Paper presented at the annual meeting of the American Educational Research Association, San Antonio, TX, April/May 2017.

Bielik, Tom, Israel Touitou, and Joseph Krajcik. "Crafting Assessments for Measuring Student Learning in Project-Based Science." Paper presented at the annual meeting of NARST, Atlanta, GA, March 2018.

Blackwell, Lisa, Kali Trzesniewski, and Carol Dweck. "Implicit Theories of Intelligence Predict Achievement across an Adolescent Transition: A Longitudinal Study and an Intervention." *Child Development* 78, no. 1 (January/February 2007): 246–263.

Bloom, Howard S. "Randomizing Groups to Evaluate Place-Based Programs." In *Learning More from Social Experiments: Evolving Analytic Approaches,* edited by Howard S. Bloom, 115–172. New York: Russell Sage, 2005.

Blumenfeld, Phyllis C., Elliott Soloway, Ronald W. Marx, Joseph S. Krajcik, Mark Guzdial, and Annemarie Palincsar. "Motivating Project-Based Learning: Sustaining the Doing, Supporting the Learning." *Educational Psychologist* 26, nos. 3–4 (1991): 369–398.

Boss, Suzie, with John Larmer. *Project Based Teaching: How to Create Rigorous and Engaging Learning Experiences.* Alexandria, VA: ASCD, 2018.

Bransford, John, Ann Brown, and Rodney Cocking. *How People Learn: Brain, Mind, Experience, and School.* Expanded edition. Washington, DC: National Academies Press, 2000.

Bryk, Anthony S. "Support a Science of Performance Improvement." *Phi Delta Kappan* (April 2009): 597–600.

Bryk, Anthony S., Louis Gomez, Alicia Grunow, and Paul LeMahieu. *Learning to Improve: How America's Schools Can Get Better at Getting Better.* Cambridge, MA: Harvard University Press, 2015.

Bryk, Anthony S., and Barbara Schneider. *Trust in Schools: A Core Resource for Improvement.* New York: Russell Sage Foundation, 2002.

Cheung, Alan, Robert E. Slavin, Cynthia Lake, and Elizabeth Kim. "Effective Secondary Science Programs: A Best-Evidence Synthesis." *Journal of Research in Science Teaching* 54, no. 1 (2016): 1–24. doi:10.1002/tea.21338.

Chopyak, Christine, and Rodger Bybee. *Instructional Materials and Implementation of Next Generation Science Standards: Demand, Supply, and Strategic Opportunities.* New York: Carnegie Corporation of New York, 2017.

Chow, Angela, and Katariina Salmela-Aro. "Task Values across Subject Domains: A Gender Comparison Using a Person-Centered Approach." *International Journal of Behavioral Development* 35, no. 3 (May 2011): 202–209. doi:10.1177/0165025411398184.

Cobb, Paul, Kara Jackson, Erin Henrick, Thomas M. Smith, and the MIST Team. *Systems for Instructional Improvement: Creating Coherence from the Classroom to the District Office.* Cambridge, MA: Harvard University Press, 2018.

Condliffe, Barbara, Janet Quint, Mary Visher, Michael Bangser, Sonia Drohojowska, Larissa Saco, and Elizabeth Nelson. "Project-Based Learning: A Literature Review." Working paper, MDRC, New York, 2017.

Corso, Michael J., Matthew J. Bundick, Russell J. Quaglia, and Dawn E. Haywood. "Where Student, Teacher, and Content Meet: Student Engagement in the Secondary School Classroom." *American Secondary Education* 41, no. 3 (Fall 2013): 50–61.

Csikszentmihalyi, Mihaly. *Flow: The Psychology of Optimal Experience.* New York: Harper Perennial, 1990.

Csikszentmihalyi, Mihaly, and Barbara Schneider. *Becoming Adult: How Teenagers Prepare for the World of Work.* New York: Basic Books, 2000.

Damelin, Daniel, Joseph Krajcik, Cynthia McIntyre, and Tom Bielik. "Students Making System Models: An Accessible Approach." *Science Scope* 40, no. 5 (2017): 78–82.

Darling-Hammond, Linda, and Ann Lieberman, eds. *Teacher Education around the World: Changing Policies and Practices.* New York: Routledge, 2012.

Design-Based Research Collective. "Design-Based Research: An Emerging Paradigm for Educational Inquiry." *Educational Researcher* 32, no. 1 (2002): 5–8.

Dewey, John. *Experience and Education.* New York: McMillan, 1938.

————. *School and Society*. Chicago: University of Chicago Press, 1907.

Dewey, John, and Albion Woodbury Small. *My Pedagogic Creed: No. 25*. New York: E. L. Kellogg & Company, 1897.

Duckworth, Angela. *Grit: The Power of Passion and Perseverance*. New York: Simon & Schuster, 2016.

Dumont, Hanna, David Istance, and Francisco Benavides, eds. *The Nature of Learning Using Research to Inspire Practice*. Paris: OECD Publishing, 2010.

Dweck, Carol. *Mindset: The New Psychology of Success*. New York: Random House, 2006.

Edelson, Daniel C., and Brian J. Reiser. "Making Authentic Practices Accessible to Learners: Design Challenges and Strategies." In *The Cambridge Handbook of the Learning Sciences*, edited by R. Keith Sawyer, 335–354. New York: Cambridge University Press, 2006.

European Commission. *Horizon 2020: Work Programme 2016–2017; European Commission Decision C(2017)2468*. Brussels: European Commission, April 24, 2017.

————. *Horizon 2020: Work Programme 2016–2017, Science with and for Society*. Brussels: European Commission, 2016.

Fishman, Barry J., William R. Penul, Anna-Ruth Allen, Britte Haugan Cheng, and Nora Sabelli. "Design-Based Implementation Research: An Emerging Model for Transforming the Relationship of Research and Practice." In *National Society for the Study of Education*, vol. 112, edited by Barry J. Fishman and William R. Penuel, 136–156. New York: Columbia Teachers' College, 2013.

FMEC (Finnish Ministry of Education and Culture). *Development Programme for Teachers' Pre- and In-Service Education*. Helsinki: Finnish Ministry of Education, 2016.

————. *Kiuru: Broad-Based Project to Develop Future Primary and Secondary Education*. Helsinki: Finnish Ministry of Education, 2014.

————. *Tulevaisuuden lukio: Valtakunnalliset tavoitteet ja tuntijako* (Future upper secondary school: National aims and allocation of lesson hours). Helsinki: Finnish Ministry of Education and Culture, 2013. http://minedu.fi/OPM/Julkaisut/2013/Tulevaisuuden _lukio.html.

FNBE (Finnish National Board of Education). *National Core Curriculum for Basic Education*. Helsinki: Finnish National Board of Education, 2014.

————. *The National Core Curriculum for Upper Secondary Education*. Helsinki: Finnish National Board of Education, 2015.

Fredricks, Jennifer A., Phyllis C. Blumenfeld, and Alison Paris. "School Engagement: Potential of the Concept, State of the Evidence." *Review of Educational Research* 74, no. 1 (Spring 2004): 59–109.

Fredricks, Jennifer A., and Wendy McColskey. "The Measurement of Student Engagement: A Comparative Analysis of Various Methods and Student Self-Report Instruments." In *Handbook of Research on Student Engagement*, edited by Sandra L. Christenson, Amy L. Reschly, and Cathy Wylie, 763–782. New York: Springer Science, 2012.

Funk, Cary, and Meg Hefferon. "As the Need for Highly Trained Scientists Grows, A Look at Why People Choose These Careers." Pew Research Center, October 24, 2016. http:// www.pewresearch.org/fact-tank/2016/10/24/as-the-need-for-highly-trained-scientists -grows-a-look-at-why-people-choose-these-careers.

Funk, Cary, and Lee Rainie. "Public and Scientists' Views on Science and Society." Pew Research Center: Internet & Technology, January 29, 2015. http://www.pewinternet.org/2015.01 /29/public-and-scientists-views-on-science-and-society.

Gago, José M., John Ziman, Paul Caro, Costas Constantinou, Graham Davies, Ilka Parchmann, Miia Rannikmäe, and Svein Sjøberg. *Europe Needs More Scientists: Report by the High Level Group on Increasing Human Resources for Science and Technology.* Brussels: European Commission, 2004. http://europa.eu.int/comm/research/conferences/2004 /sciprof/pdf/final en.pdf.

Graham, Karen J., Lara M. Gengarelly, Barbara A. Hopkins, and Melissa Lombard. *Dive In! Immersion in Science Practices for High School Students.* Arlington, VA: National Science Teachers Association, 2017.

Harris, Christopher J., Joseph S. Krajcik, James W. Pellegrino, and Kevin W. McElhaney. *Constructing Assessment Tasks That Blend Disciplinary Core Ideas, Crosscutting Concepts, and Science Practices for Classroom Formative Applications.* Menlo Park, CA: SRI International, 2016.

Hedges, Larry V. "Challenges in Building Usable Knowledge in Education." *Journal of Research on Educational Effectiveness* 11, no. 1 (2018): 1–21. doi:10.1080/19345747.2017.1375583.

Hektner, Joel, Jennifer A. Schmidt, and Mihaly Csikszentmihalyi. *Experience Sampling Method: Measuring the Quality of Everyday Life.* Thousand Oaks, CA: SAGE Publications, 2011.

Hidi, Suzanne, and Ann Renniger. "The Four Phase Model of Interest Development." *Educational Psychologist* 41, no. 2 (2006): 111–127.

Hietin, Liana. "2 in 5 High Schools Don't Offer Physics, Analysis Finds." *Education Week,* August 23, 2016.

Hinojosa, Trisha, Amie Rapaport, Andrew Jaciw, Christina LiCalsi, and Jenna Zacamy. *Exploring the Foundations of the Future STEM Workforce: K–12 Indicators of Postsecondary STEM Success.* Washington, DC: U.S. Department of Education, 2016.

Horner, Robert H., and Samuel L. Odom. "Constructing Single-Case Research Designs: Logic." In *Advances,* edited by Thomas R. Kratochwill and Joel R. Levin, 27–51. Washington, DC: American Psychological Association, 2014.

Immordino-Yang, Mary Helen. *Emotions, Learning, and the Brain: Exploring the Educational Implications of Affective Neuroscience.* New York: W.W. Norton, 2015.

Inkinen, Janna, Christopher Klager, Barbara Schneider, Kalle Juuti, Joseph Krajcik, Jari Lavonen, and Katariina Salmela-Aro. "Science Classroom Activities and Student Situational Engagement." *International Journal of Science Education* 41, no. 3 (2019): 316–329.

Jacob, Brian, Susan Dynarksi, Kenneth Frank, and Barbara Schneider. "Are Expectations Alone Enough? Estimating the Effect of Mandatory College-Prep Curriculum in Michigan." *Education Evaluation and Policy Analysis* 39, no. 2 (2017): 333–360.

Jones, Stephanie M., and Emily J. Doolittle. "Social and Emotional Learning." In *The Future of Children,* vol. 27, no. 1, edited by Stephanie M. Jones and Emily J. Doolittle, 3–12. Princeton, NJ: Princeton-Brookings, 2017.

Juuti, Kalle, and Jari Lavonen. "How Teaching Practices Are Connected to Student Intention to Enroll in Upper Secondary School Physics Courses." *Research in Science & Technological Education* 34, no. 2 (2016): 204–218.

Juuti, Kalle, Jari Lavonen, and Veijo Meisalo. "Pragmatic Design-Based Research—Designing as a Shared Activity of Teachers and Researches." In *Iterative Design of Teaching-Learning Sequences: Introducing the Science of Materials in European Schools,* edited by Dimitris Psillos and Petros Kariotoglou, 35–46. Heidelberg: Springer Dordrecht, 2015.

Kanter, David E. "Doing the Project and Learning the Content: Designing Project-Based Science Curricula for Meaningful Understanding." *Science Education* 94, no. 3 (2010): 525–551.

Kastberg, David, Jessica Ying Chan, Gordon Murray, and Patrick Gonzales. *Performance of US 15-Year-Old Students in Science, Reading, and Mathematics Literacy in an International Context: First Look at PISA 2015.* Washington, DC: U.S. Department of Education, December 2016.

Kennedy, Craig H. *Single-Case Designs for Educational Research.* Boston: Allyn & Bacon, 2005.

Kilpatrick, William H. "The Project Method." *Teachers College Record* 19, no. 4 (1918): 319–335. http://www.tcrecord.org.

———. "Project Teaching." *General Science Quarterly* 1, no. 2 (1917): 67–72.

Klager, Christopher. "Project-Based Learning in Science: A Meta-Analysis of Science Achievement Effects." Working paper, Michigan State University, 2017.

Klager, Christopher, Richard Chester, and Israel Touitou. "Social and Emotional Experiences of Students Using an Online Modeling Tool." Paper presented at the annual meeting of NARST, Atlanta, GA, March 2018.

Klager, Christopher, and Janna Inkinen. "Socio-emotional Experiences of Students in Science and Other Academic Classes." Paper presented at the annual meeting of the American Education Research Association, Washington, D.C., April 2016.

Klager, Christopher, and Barbara Schneider. "Strategies for Evaluating Curricular Interventions Using the Experience Sampling Method." Poster presented at the annual meeting of the Society for Research on Educational Effectiveness, Washington, DC, February/March 2018.

———. "Enhancing Imagination and Problem-Solving Using Project-Based Learning." Working paper, Michigan State University, 2017.

Klager, Christopher, Barbara Schneider, Joseph Krajcik, Jari Lavonen, and Katariina Salmela-Aro. "Creativity in a Project-Based Physics and Chemistry Intervention." Paper presented at the annual meeting of NARST, San Antonio, TX, April 2017.

Klager, Christopher, Barbara Schneider, and Katariina Salmela-Aro. "Enhancing Imagination and Problem-Solving in Science." Paper presented at the annual meeting of the American Education Research Association, New York, April 2018.

Knoll, Michael. "'I Had Made a Mistake': William H. Kilpatrick and the Project Method." *Teachers College Record* 114, no. 2 (February 2012): 1–45.

Korhonen, Tiina, and Jari Lavonen. "Crossing School–Family Boundaries through the Use of Technology." In *Crossing Boundaries for Learning—Through Technology and Human Efforts,* edited by Hannele Niemi, Jari Multisilta, and Erika Löfström, 48–66. Helsinki: CICERO Learning, 2014.

Krajcik, Joseph. "Project-Based Science: Engaging Students in Three-Dimensional Learning." *Science Teacher* 82, no. 1 (2015): 25.

Krajcik, Joseph, and Phyllis C. Blumenfeld. "Project-Based Learning." In *The Cambridge Handbook of the Learning Sciences,* edited by R. Keith Sawyer, 317–334. New York: Cambridge University Press, 2005.

Krajcik, Joseph, Susan Codere, Chanyah Dahsah, Renee Bayer, and Kongju Mun. "Planning Instruction to Meet the Intent of the Next Generation Science Standards." *Journal of Science Teacher Education* 25, no. 2 (2014): 157–175. doi 10.1007/s10972–014–9383–2.

Krajcik, Joseph, and Charlene Czerniak. *Teaching Science in Elementary and Middle School: A Project-Based Approach,* 4th ed. London: Taylor & Francis, 2013.

Krajcik, Joseph, and Joi Merritt. "Engaging Students in Scientific Practices: What Does Constructing and Revising Models Look Like in the Science Classroom? Understanding 'A Framework for K–12 Science Education.'" *Science Teacher* 79, no. 3 (2012): 38–41.

Krajcik, Joseph, and Namsoo Shin. "Project-Based Learning." In *The Cambridge Handbook of the Learning Sciences,* 2nd ed., edited by R. Keith Sawyer, 275–297. New York: Cambridge University Press, 2015.

Kratochwill, Thomas R., ed. *Single Subject Research: Strategies for Evaluating Change.* New York: Academic Press, 1978.

Kratochwill, Thomas R., and Joel R. Levin. "Introduction: An Overview of Single-Case Intervention Research." In *Single-Case Intervention Research: Methodological and Statistical Advances,* edited by Thomas R. Kratochwill and Joel R. Levin, 3–23. Washington, DC: American Psychological Association, 2014.

Lane, Justin D., and David L. Gast. "Visual Analysis in Single Case Experimental Design Studies: Brief Review and Guidelines." *Neuropsychological Rehabilitation* 24, nos. 3–4 (2014): 445–463.

Larmer, John, John Mergendoller, and Suzie Boss. *Setting the Standard for Project Based Learning: A Proven Approach to Rigorous Classroom Instruction.* Alexandria, VA: ASCD, 2015.

Lavonen, Jari. "Building Blocks for High-Quality Science Education: Reflections Based on Finnish Experiences." *LUMAT* 1, no. 3 (2013): 299–313.

———. "Educating Professional Teachers in Finland through the Continuous Improvement of Teacher Education Programmes." In *Contemporary Pedagogies in Teacher Education and Development,* edited by Yehudith Weinberger and Zipora Libman, 3–17. London: IntechOpen, 2018. http://dx.doi.org/10.5772/intechopen.77979.

———. "Educating Professional Teachers through the Master's Level Teacher Education Programme in Finland." *Bordón* 68, no. 2 (2016): 51–68.

———. "The Influence of an International Professional Development Project for the Design of Engaging Secondary Science Teaching in Finland." Paper presented at the 25th annual meeting of the Southern African Association of Researchers in Mathematics Science and Technology Education, Johannesburg, January 2017.

———. "National Science Education Standards and Assessment in Finland." In *Making It Comparable: Standards in Science Education,* edited by David J. Waddington, Peter Nentwig, and Sascha Schaze, 101–126. Berlin: Waxman, 2007.

Lavonen, Jari, and Kalle Juuti. "Evaluating Learning of Conceptual, Procedural, and Epistemic Knowledge in a Project-Based Learning Unit." Paper presented at the annual meeting of NARST, Atlanta, GA, March 2018.

———. "Science at Finnish Compulsory School." In *The Miracle of Education: The Principles and Practices of Teaching and Learning in Finnish Schools,* edited by Hannele Niemi, Auli Toom, and Arto Kallioniemi, 131–147. Rotterdam: Sense Publishers, 2012.

Lavonen, Jari, Kalle Juuti, Maija Aksela, and Veijo Meisalo. "A Professional Development Project for Improving the Use of ICT in Science Teaching." *Technology, Pedagogy and Education* 15, no. 2 (2006): 159–174. doi:10.1080/14759390600769144.

Lavonen, Jari, Kalle Juuti, Anna Uitto, Veijo Meisalo, and Reijo Byman. "Attractiveness of Science Education in the Finnish Comprehensive School." In *Research Findings on Young People's Perceptions of Technology and Science Education,* edited by Anneli Manninen, Kirsti Miettinen, and Kati Kiviniemi, 5–30. Helsinki: Technology Industries of Finland, 2005.

Lavonen, Jari, and Seppo Laaksonen. "Context of Teaching and Learning School Science in Finland: Reflections on PISA 2006 Results." *Journal of Research in Science Teaching* 46, no. 8 (2009): 922–944.

Lee, Okhee, and Cory A. Buxton. *Diversity and Equity in Science Education: Theory, Research, and Practice.* New York: Teachers College Press, 2010.

Linnansaari, Janna, Jaana Viljaranta, Jari Lavonen, Barbara Schneider, and Katariina Salmela-Aro. "Finnish Students' Engagement in Science Lessons." *Nordic Studies in Science Education* 11, no. 2 (2015): 192–206.

Linnenbrink-Garcia, Lisa, Erika A. Patall, and Reinhard Pekrun. "Adaptive Motivation and Emotion in Education: Research and Principles for Instructional Design." *Policy Insights from the Behavioral and Brain Sciences* 3, no. 2 (2016): 228–236.

Little, Brian R., Katariina Salmela-Aro, and Susan D. Phillips. *Personal Project Pursuit: Goals, Action and Human Flourishing.* Hillsdale, NJ: Lawrence Erlbaum, 2007.

Llewellyn, Douglas. *Teaching High School Science through Inquiry and Argumentation,* 2nd ed. Thousand Oaks, CA: Corwin, 2013.

Lucas, Bill, Guy Claxton, and Ellen Spencer. "Progression in Student Creativity in School: First Steps towards New Forms of Formative Assessments." OECD Education Working Paper no. 86, 2013. http://dx.doi.org/10.1787/5k4dp59msdwk-en.

Luft, Julie A., and Peter W. Hewson. "Research on Teacher Professional Development Programs in Science." In *Handbook of Research in Science Education,* 2nd ed., edited by Sandra K. Abell and Norman G. Lederman, 889–909. New York: Taylor Francis, 2014.

Markham, Thom, John Larmer, and Jason Rabitz. *Project-Based Learning Handbook: A Guide to Standards-Focused Project Based Learning for Middle and High School Teachers.* Novato, CA: Buck Institute for Education, 2003.

McNeill, Katherine L., Rebecca Katsh-Singer, and Pam Pelletier. "Assessing Science Practices: Moving Your Class along a Continuum." *Science Scope* 39, no. 4 (2015): 21–28.

McNeill, Katherine, and Joseph Krajcik. *Supporting Grade 5–8 Students in Constructing Explanations in Science: The Claim, Evidence, and Reasoning Framework for Talk and Writing.* New York: Allyn & Bacon, 2012.

Mergendoller, John R., Nan L. Maxwell, and Yolanda Bellisimo. "The Effectiveness of Problem-Based Instruction: A Comparative Study of Instructional Methods and Student Characteristics." *Interdisciplinary Journal of Problem-Based Learning* 1, no. 2 (2006): 49–69.

Miller, Jon. "The Conceptualization and Measurement of Civic Scientific Literacy for the Twenty-First Century." In *Science and the Educated American: A Core Component of Liberal Education,* edited by Jerrold Meinwald and John G. Hildebrand, 241–255. Cambridge, MA: American Academy of Arts and Sciences, 2010.

Moeller, Julia, Katariina Salmela-Aro, Jari Lavonen, and Barbara Schneider. "Does Anxiety in Math and Science Classrooms Impair Motivations? Gender Differences beyond the Mean Level." *International Journal of Gender, Science, and Technology* 7, no. 2 (June 2015): 229–254.

Mullis, Ina V. S., Michael O. Martin, Pierre Foy, and Martin Hooper. *TIMSS Advanced 2015 International Results in Advanced Mathematics and Physics.* Boston: Boston College, 2016.

Muthén, Linda K., and Bengt Muthén. *Mplus User's Guide.* Los Angeles: Muthén & Muthén, 2015.

National Academies of Sciences, Engineering, and Medicine. *How People Learn II: Learners, Contexts, and Cultures.* Washington, DC: National Academies Press, 2018. https://doi.org/10.17226/24783.

———. *Rising above the Gathering Storm: Energizing and Employing America for a Brighter Economic Future.* Washington, DC: National Academies Press, 2007.

NCES (National Center for Education Statistics). *Schools and Staffing Survey.* Washington, DC: National Center for Education Statistics, 2016.

NGSS (Next Generation Science Standards) Lead States. *Next Generation Science Standards: For States, By States.* Washington, DC: National Academies Press, 2013.

Niemi, Hannele. "Educating Student Teachers to Become High Quality Professionals: A Finnish Case." *CEPS Journal* 1, no. 1 (2011): 43–66.

Niemi, Hannele, Auli Toom, and Arto Kallioniemi. *The Miracle of Education: The Principles and Practices of Teaching and Learning in Finnish Schools.* Rotterdam: Sense Publishers, 2012.

No Child Left Behind Act of 2001. Pub L. No. 107–110, 20 U.S.C. § 6319 (2002).

NRC (National Research Council). *A Framework for K–12 Science Education: Practices, Crosscutting Concepts, and Core Ideas.* Washington, DC: National Academies Press, 2012.

———. *Guide to Implementing the Next Generation Science Standards.* Washington, DC: National Academies Press, 2015.

NSB (National Science Board). *Science & Engineering Indicators 2018.* January 2018. https://www.nsf.gov/statistics/2018/nsb20181/.

NSF (National Science Foundation). *Preparing the Next Generation of STEM Innovators: Identifying and Developing Our Human Capital.* Washington, DC: National Science Foundation, 2010.

NSTA (National Science Teachers Association). *NGSS@NSTA: Stem Starts Here.* http://www.ngss.nsta.org.

OECD (Organisation for Economic Co-operation and Development). *PISA 2006,* vol 1: *Science Competencies for Tomorrow's World.* Paris: OECD Publishing, 2007. http://dx.doi.org/10.1787/9789264040014-en.

———. *PISA 2012 Results in Focus: What 15-Year-Olds Know and What They Can Do with What They Know.* Paris: OECD Publishing, 2014.

———. *PISA 2015 Draft Science Framework.* Paris: OECD Publishing, 2013.

———. *PISA 2015 Results,* vol. 1: *Excellence and Equity in Education.* Paris: OECD Publishing, 2016. http://dx.doi.org/10.1787/9789264266490-en.

———. *PISA 2015 Results,* vol. 3: *Students' Well-Being.* Paris: OECD Publishing, 2017.

———. *TALIS 2013 Results: An International Perspective on Teaching and Learning.* Paris: OECD Publishing, 2014.

OECD (Organisation for Economic Co-operation and Development) Global Science Forum. *Evolution of Student Interest in Science and Technology Studies.* Policy report. Paris: OECD Publishing, 2006.

Oikkonen, Juha, Jari Lavonen, Heidi Krzywacki-Vainio, Maija Aksela, Leena Krokfors, and Heimo Saarikko. "Pre-Service Teacher Education in Chemistry, Mathematics, and Physics." In *How Finns Learn Mathematics and Science,* edited by Erkki Pehkonen, Maija Ahtee, and Jari Lavonen, 49–68. Rotterdam: Sense Publishers, 2007.

Osborne, Jonathan, and Justin Dillon. *Science Education in Europe: Critical Reflections.* London: Nuffield Foundation, January 2008.

Osborne, Jonathan, Shirley Simon, and Sue Collins. "Attitudes towards Science: A Review of the Literature and Its Implications." *International Journal of Science Education* 25, no. 9 (2003): 1049–1079.

Peek-Brown, Deborah, Kellie Finnie, Joseph S. Krajcik, and Tom Bielik. "Using Artifacts Developed in Project-Based Learning Classrooms as Evidence of 3-Dimensional Learning." Paper presented at the annual meeting of NARST, Atlanta, GA, March 2018.

Perez-Felkner, Lara, Sarah-Kathryn McDonald, Barbara Schneider, and Erin Grogan. "Female and Male Adolescents' Subjective Orientations to Mathematics and Their Influence on

Postsecondary Majors." *Developmental Psychologist* 48, no. 6 (2012): 1658–1673. doi:10.1037/a0027020.

Pietarinen, Janne, Kirsi Pyhältö, Tiina Soini, and Katariina Salmela-Aro. "Reducing Teacher Burnout: A Socio-Contextual Approach." *Teaching and Teacher Education* 35 (October 2013): 62–72. doi:10.1016/j.tate.2013.05.003.

———. "Validity and Reliability of the Socio-Contextual Teacher Burnout Inventory (STBI)." *Psychology* 4, no. 1 (January 2013): 73–82. doi:10.4236/psych.2013.41010.

Polman, Joseph L. *Designing Project-Based Science, Connecting Learners through Guided Inquiry.* New York: Teachers College Press, 2000.

Provasnik, Stephen, Lydia Malley, Maria Stephens, Katherine Landeros, Robert Perkins, and Judy H. Tang. *Highlights from TIMSS and TIMSS Advanced 2015: Mathematics and Science Achievement of US Students in Grades 4 and 8 and in Advanced Courses at the End of High School in an International Context (NCES 2017–002).* Washington, DC: U.S. Department of Education, National Center for Education Statistics, 2016.

Pyhältö, Kirsi, Janne Pietarinen, and Katariina Salmela-Aro. "Teacher–Working-Environment Fit as a Framework for Burnout Experienced by Finnish Teachers." *Teaching and Teacher Education* 27, no. 7 (2011): 1101–1110. doi:10.1016/j.tate.2011.05.006.

Quin, Daniel. "Longitudinal and Contextual Associations between Teacher–Student Relationships and Student Engagement: A Systematic Review." *Review of Educational Research* 87, no. 2 (April 2017): 345–387.

Raudenbush, Stephen W., and Anthony S. Bryk. *Hierarchical Linear Models,* 2nd ed. Thousand Oaks, CA: SAGE, 2002.

Rindskopf, David M., and John M. Ferron. "Using Multilevel Models to Analyze Single-Case Design Data," in *Single-Case Intervention Research: Methodological and Statistical Advances,* edited by Thomas R. Kratochwill and Joel R. Levin, 221–246. Washington, DC: American Psychological Association, 2014.

Rocard, Michel, Peter Csermely, Doris Jorde, Dieter Lenzen, Harriet Walberg-Henriksson, and Valerie Hemmo. *Science Education Now: A Renewed Pedagogy for the Future of Europe.* Brussels: European Commission, 2006. http://ec.europa.ed/research/science-society /documentlibrary/pdf/06/report-ocard-on-science-education.

Rushton, Gregory T., David Rosengrant, Andrew Dewar, Lisa Shah, Herman E. Ray, Keith Sheppard, and Lynn Watanabe. "Towards a High Quality High School Workforce: A Longitudinal, Demographic Analysis of US Public School Physics Teachers." *Physical Review Physics Education Research* 13, no. 020122 (October 23, 2017).

Ryan, Richard M., and Edward L. Deci. *Self-Determination Theory: Basic Psychological Needs in Motivation, Development, and Wellness.* New York: Guilford Press, 2017.

Sahlberg, Pasi. *Finnish Lessons: What Can the World Learn from Educational Change in Finland?* New York: Teachers College Press, 2011.

———. *Finnish Lessons 2.0: What Can the World Learn from Educational Change in Finland?* 2nd ed. New York: Teachers College Press, 2015.

Salmela-Aro, Katariina. "Dark and Bright Sides of Thriving—School Burnout and Engagement in the Finnish Context." *European Journal of Developmental Psychology* 14, no. 3 (2017): 337–349. doi:10.1080/17405629.2016.1207517.

Salmela-Aro, Katariina. "Personal Goals and Well-Being during Critical Life Transitions: The Four C's—Channelling, Choice, Co-Agency and Compensation." *Advances in Life Course Research* 14, nos. 1–2 (2009): 63–73. doi:10.1016/j.alcr.2009.03.003.

Salmela-Aro, Katariina, Noona Kiuru, Esko Leskinen, and Jari-Erik Nurmi. "School Burnout
    Inventory (SBI): Reliability and Validity." *European Journal of Psychological Assessment* 8,
    no. 1 (2009): 60–67.
Salmela-Aro, Katariina, Jukka Marjanen, Barbara Schneider, Jari Lavonen, and Christopher
    Klager. "Does It Help to Have 'Sisu'? The Role of Situational Grit and Challenge in
    Science Situations among Finnish and US Secondary Students." Working paper,
    University of Helsinki, 2018.
Salmela-Aro, Katariina, Julia Moeller, Barbara Schneider, Justina Spicer, and Jari Lavonen.
    "Integrating the Light and Dark Sides of Student Engagement Using Person-Oriented
    and Situation-Specific Approaches." *Learning and Instruction* 43 (2016): 61–70. doi:10
    .1016/j.learninstruc.2016.01.001.
Salmela-Aro, Katariina, and Sanna Read. "Study Engagement and Burnout Profiles among
    Finnish Higher Education Students." *Burnout Research* 7 (2017): 21–28. doi:10.1016
    /j.burn.2017.11.001.
Salmela-Aro, Katariina, and Katja Upadyaya. "School Burnout and Engagement in the Context
    of Demands-Resources Model." *British Journal of Educational Psychology* 84, no. 4
    (2014): 137–151.
Salmela-Aro, Katariina, Katja Upadyaya, Kai Hakkarainen, Kirsti Lonka, and Kimmo Albo.
    "The Dark Side of Internet Use: Two Longitudinal Studies of Excessive Internet Use,
    Depressive Symptoms, School Burnout and Engagement among Finnish Early and Late
    Adolescents." *Journal of Youth Adolescence* 46, no. 2 (May 2016): 343–357. doi:10.10007
    /s10964–0494–2.
Savin Baden, Maggi, and Claire Howell Major. *Foundations of Problem-Based Learning.* London:
    Open University Press/McGraw-Hill Education, 2009.
Schmidt, Jennifer A., Joshua M. Rosenberg, and Patrick N. Beymer. "A Person-in-Context Ap-
    proach to Student Engagement in Science: Examining Learning Activities and Choice."
    Journal of Research in Science Teaching 55, no. 1 (2018): 19–43.
Schneider, Barbara, Martin Carnoy, Jeremy Kilpatrick, William H. Schmidt, and Richard J.
    Savelson. *Estimating Causal Effects Using Experimental and Observational Designs.* Wash-
    ington, DC: American Educational Research Association, 2007.
Schneider, Barbara, I-Chien Chen, and Christopher Klager. "Gender Equity and the Allocation of
    Parent Resources: A Forty Year Analysis." Paper presented at an invited symposium at
    the University of Helsinki, May 2018.
Schneider, Barbara, Kenneth Frank, I-Chien Chen, Venessa Keesler, and Joseph Martineau. "The
    Impact of Being Labeled as a Persistently Lowest Achieving School: Regression Discon-
    tinuity Evidence on School Labeling." *American Journal of Education* 123, no. 4 (August
    2017): 585–613. doi:10.1086/692665.
Schneider, Barbara, Christopher Klager, I-Chien Chen, and Jason Burns. "Transitioning into
    Adulthood: Striking a Balance between Support and Independence." *Policy Insights from
    Behavioral and Brain Sciences* 3, no. 3 (January 2016): 106–113. doi:10.1177/23727322156
    24932.
Schneider, Barbara, Joseph Krajcik, Christopher Klager, and Israel Touitou. "Developing Three-
    Dimensional Assessment Tasks and Mapping Their Relation to Existing Testing Frame-
    works." Paper presented at the annual meeting of NARST, Atlanta, GA, March 2018.
Schneider, Barbara, Joseph Krajcik, Jari Lavonen, Katariina Salmela-Aro, Michael Broda, Justina
    Spicer, Justin Bruner, Julia Moeller, Janna Linnansaari, Kalle Juuti, and Jaana Viljaranta.

"Investigating Optimal Learning Moments in US and Finnish Classrooms." *Journal of Research in Science Teaching* 53, no. 3 (December 2015): 400–421. doi:10.1002/tea.21306.

Schneider, Barbara, Carolina Milesi, Lara Perez-Felkner, Kevin Brown, and Iliya Gutin. "Does the Gender Gap in STEM Majors Vary by Field and Institutional Selectivity?" *Teachers College Record* 20 (2015).

Schneider, Barbara, and David Stevenson. *The Ambitious Generation: America's Teenagers, Motivated but Directionless.* New Haven: Yale University Press, 1999.

Shadish, William R., Thomas D. Cook, and Donald T. Campbell. *Experimental and Quasi-Experimental Designs for Generalized Causal Inference.* Belmont, CA: Wadsworth Cengage Learning, 2002.

Shalley, Christina, Michael Hitt, and Jing Zhou, eds. *The Oxford Handbook of Creativity, Innovation, and Entrepreneurship.* New York: Oxford University Press, 2015.

Shernoff, David J., Mihaly Csikszentmihalyi, Barbara Schneider, and Elisa Steele Shernoff. "Student Engagement in High School Classrooms from the Perspective of Flow Theory." *School Psychology Quarterly* 18, no. 2 (2003): 158–176.

Shernoff, David J., Shaunti Knauth, and Eleni Makris. "The Quality of Classroom Experiences." In *Becoming Adult: How Teenagers Prepare for the World of Work,* edited by Mihaly Csikszentmihalyi and Barbara Schneider, 141–164. New York: Basic Books, 2000.

Shumow, Lee, Jennifer Schmidt, and Hayal Kacker. "Adolescents' Experience Doing Homework: Associations among Context, Quality of Experience, and Outcomes." *School Community Journal* 18, no. 2 (2008).

Sjøberg, Svein, and Camilla Schreiner. *The ROSE Project: An Overview and Key Findings.* Oslo: University of Oslo, March 2010.

Snow, Catherine E., and Kenne A. Dibner, eds. *Science Literacy: Concepts, Contexts, and Consequences.* Washington, DC: National Academies Press, 2016.

Spicer, Justina, Barbara Schneider, Katariina Salmela-Aro, and Julia Moeller. "The Conceptualization and Measurement of Student Engagement in Science: A Cross-Cultural Examination from Finland and the United States." In *Global Perspectives on Education Research,* edited by Lori Diane Hill and Felice J. Levine, 227–249. New York: Routledge, 2018.

Symonds, Jennifer, Ingrid Schoon, and Katariina Salmela-Aro. "Developmental Trajectories of Emotional Disengagement from Schoolwork and Their Longitudinal Associations in England." *British Educational Research Journal* 42, no. 6 (September 2016): 993–1022. doi: 10.1002/berj.3243.

TeachingWorks. "Deborah Lowenberg Ball—Director." www.teachingworks.org.

Thomas, John A. *A Review of Research on Project-Based Learning.* San Rafael, CA: The Autodesk Foundation, March 2000.

Touitou, Israel, Stephen Barry, Tom Bielik, Barbara Schneider, and Joseph Krajcik. "The Activity Summary Board: Adding a Visual Reminder to Enhance a Project-Based Learning Unit." *Science Teacher* (March 2018): 30–35.

Touitou, Israel, and Joseph Krajcik. "Developing and Measuring Optimal Learning Environments through Project-Based Learning." Paper presented at the annual meeting of the American Education Research Association, New York, April 2018.

Touitou, Israel, Joseph S. Krajcik, Barbara Schneider, Christopher Klager, and Tom Bielik. "Effects of Project-Based Learning on Student Performance: A Simulation Study." Paper presented at the annual meeting of NARST, Atlanta, GA, March 2018.

Tuominen-Soni, Heta, and Katariina Salmela-Aro. "Schoolwork Engagement and Burnout among Finnish High School Students and Young Adults: Profiles, Progressions, and Educational Outcomes." *Developmental Psychology* 50, no. 3 (2014): 649–662.

Upadyaya, Katja, Katariina Salmela-Aro, Barbara Schneider, Jari Lavonen, Joseph Krajcik, and Christopher Klager. "Associations between Students' Task Values and Emotional Learning Experiences in Science Situations among Finnish and US Secondary School Students." Working paper, University of Helsinki, 2018.

Vincent-Lancrin, Stéphan. *Teaching, Assessing, and Learning Creative and Critical Thinking Skills in Education.* Paris: OECD Centre for Educational Research and Innovation, 2018. http://www.oecd.org/education/ceri/assessingprogressionincreativeandcriticalthinkingskillsineducation.htm.

Vitikka, Erja, Leena Krokfors, and Elisa Hurmerinta. "The Finnish National Core Curriculum." In *The Miracle of Education: The Principles and Practices of Teaching and Learning in Finnish Schools,* edited by Hannele Niemi, Auli Toom, and Arto Kallioniemi, 83–96. Rotterdam: Sense Publishers, 2012.

Watch Mr. Wizard. 2004. http://www.mrwizardstudios.com.

Yeager, David S. "Social and Emotional Learning Programs for Adolescents." In *The Future of Children,* vol. 27, no. 1, edited by Stephanie M. Jones and Emily J. Doolittle, 73–94. Princeton, NJ: Princeton-Brookings, 2017.

Index

Page numbers in italics indicate illustrations.

About the Authors

Barbara Schneider is the John A. Hannah University Distinguished Professor in the College of Education and Department of Sociology at Michigan State University. Professor Schneider has published multiple books and numerous articles and reports on family, the social context of schooling, and the sociology of knowledge. Past president of the American Educational Research Association, she is a fellow of the American Association for the Advancement of Science, National Academy of Education, and American Educational Research Association.

Joseph Krajcik is the Lappan-Phillips Professor of Science Education and the director of the CREATE for STEM Institute at Michigan State University. Recognized internationally for his achievements in science education, he has been a visiting professor at universities in South Korea, China, and Israel and the recipient of several awards for his contributions to science education. Professor Krajcik is a fellow of the American Association for the Advancement of Science, National Academy of Education, and the American Educational Research Association.

Jari Lavonen is professor in physics and chemistry education at the University of Helsinki and director of the National Teacher Education Forum. A leader in science and mathematics education for more than thirty years, he is the author or editor of numerous papers, articles, and books on the topic of science education and teaching. He is a Distinguished Visiting Professor at the University of Johannesburg in South Africa and has consulted widely in Norway and Peru. He is a member of the Finnish Academy of Science and Letters.

Katariina Salmela-Aro is professor of educational psychology at the University of Helsinki. Author of multiple articles, reports, and grants, she is widely known for her seminal work on school engagement, burnout, and optimal learning moments. She was previously president of the European Association in Developmental Psychology and secretary general of the International Society for the Study of Behavioral Development. She was recently awarded the Marie Curie Visiting Professorship at the University of Zürich and is a member of the Finnish Academy of Science and Letters.

Margaret J. Geller is senior scientist at the Harvard-Smithsonian Center for Astrophysics in Cambridge, Massachusetts. She is a fellow of the American Association for the Advancement of Science, American Physical Society, American Academy of Arts and Sciences, and National Academy of Sciences. She has received seven honorary degrees.